REVISE AQA GCSE
Mathematics A

REVISION GUIDE

Higher

Series Consultant: Harry Smith   Author: Harry Smith

THE REVISE AQA SERIES
Available in print or online

Online editions for all titles in the Revise AQA series are available Spring 2013.

Presented on our ActiveLearn platform, you can view the full book and customise it by adding notes, comments and weblinks.

Print editions

Mathematics A Revision Guide Higher	9781447941361
Mathematics A Revision Workbook Higher	9781447941446

Online editions

Mathematics A Revision Guide Higher	9781447941378
Mathematics A Revision Workbook Higher	9781447941514

This Revision Guide is designed to complement your classroom and home learning, and to help prepare you for the exam. It does not include all the content and skills needed for the complete course. It is designed to work in combination with Pearson's main AQA GCSE Mathematics 2010 Series.

To find out more visit:
www.pearsonschools.co.uk/aqagcsemathsrevision

ALWAYS LEARNING   PEARSON

Contents

A small bit of small print

AQA publishes Sample Assessment Material and the Specification on its website. That is the official content and this book should be used in conjunction with it. The questions in *Now try this* have been written to help you practise every topic in the book. Remember: the real exam questions may not look like this.

Target grades

Target grades are quoted in this book for some of the questions. Students targeting this grade should be aiming to get most of the marks available. Students targeting a higher grade should be aiming to get all of the marks available.

A*
A
B
C
D

Calculator skills 1

You can take a calculator into your Unit 1 exam. You need to know how to use your calculator to solve problems involving fractions and percentages.

Calculating with fractions

You can enter fractions on your calculator using the ▦ key and the arrows. For example, to work out $\frac{3}{8}$ of 52:

$$\frac{3}{8} \times 52$$
$$\frac{39}{2}$$

If you want to convert an answer on your calculator display from a fraction to a decimal you can use the S⇔D key.

Quick conversions

You can convert between fractions, decimals and percentages quickly using a calculator:

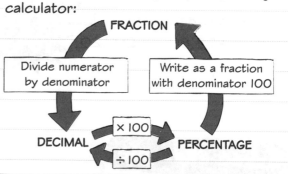

FRACTION

Divide numerator by denominator

Write as a fraction with denominator 100

× 100

DECIMAL   ÷ 100   PERCENTAGE

You can solve lots of percentage problems by working out what 1% represents. To find a percentage of an amount:

> Divide the percentage by 100
>
> ↓
>
> Multiply by the amount

Everything in red is part of the answer.

Worked example

target **D**

A company gives 3.5% of its profits to charity. In 2011 the company made profits of £470 000. How much money did the company give to charity in 2011?

(3 marks)

$3.5 \div 100 = 0.035$

$0.035 \times 470\,000 = 16\,450$

The company gave £16 450 to charity.

Worked example

In a year group of 85 students, 62 buy their lunch at school. What percentage of students buy their lunch in school? Give your answer to 1 decimal place.

(3 marks)

$62 \div 85 = 0.72941...$

$0.72941... \times 100 = 72.941...$

72.9% of students buy their lunch in school.

To write one quantity as a percentage of another:

> Divide the first quantity by the second quantity
>
> ↓
>
> Multiply your answer by 100

Always write down at least five digits from your calculator display before rounding your answer.

Now try this

1 A school has 1150 students. 54% of the students are girls. How many boys are there?

(3 marks)

2 43 of the 56 houses in a street have broadband. What percentage of the houses have broadband? Give your answer to 1 decimal place. *(3 marks)*

Percentage change 1

There are two methods that can be used to increase or decrease an amount by a percentage.

Method 1

Work out 26% of £280:

$\frac{26}{100} \times £280 = £72.80$

Subtract the decrease:

£280 − £72.80 = £207.20

Method 2

Use a multiplier:

100% + 30% = 130%

$\frac{130}{100} = 1.3$

So the multiplier for a 30% increase is 1.3:

400 g × 1.3 = 520 g

400 g PLUS 30% EXTRA

Worked example

target **D**

A football club increases the prices of its season tickets by 5.2% each year.

In 2011 a top-price season ticket cost £650.

Calculate the price of this season ticket in 2012.   *(2 marks)*

$\frac{5.2}{100} \times £650 = £33.80$

£650 + £33.80 = £683.80

When working with money, you must give your answer to 2 decimal places.

Check it!

Choose an easy percentage which is close to 5.2%.

10% of £650 is £65, so 5% is £32.50.

£650 + £32.50 = £682.50, which is close to £683.80 ✓

Calculating a percentage increase or decrease

Work out the amount of the increase or decrease

⬇

Write this as a percentage of the original amount

Was £60 Now £39

60 − 39 = 21

$\frac{21}{60} = 35\%$

This is a 35% decrease.

For a reminder about writing one quantity as a percentage of another, have a look at page 1.

A question may ask you to calculate a percentage **profit** or **loss** rather than an increase or decrease.

For a reminder about writing one quantity as a percentage of another, have a look at page 1.

Now try this

target **D**

1   A car manufacturer increases its prices by 8%.

The price of a particular model before the increase was £13 250.

What will this particular model cost after the price increase?   *(3 marks)*

target **C**

2   A TV originally cost £520.

In a sale it was priced at £340.

What was the percentage reduction in the price? Give your answer to 1 decimal place.   *(3 marks)*

Reduction means decrease. Work out the decrease as a percentage of the original price.

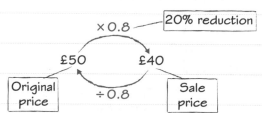

-A*
-A
-B
-C
-D

Reverse percentages and compound interest

Reverse percentages

In some questions you are given an amount AFTER a percentage change.

To find the original amount you divide by the MULTIPLIER.

×0.8 — 20% reduction

£50 → £40

Original price ÷0.8 → Sale price

Worked example

target B

In a sale, normal prices are reduced by 15%. The sale price of a pair of trainers is £75.65

Work out the normal price of the trainers.

(3 marks)

$100\% - 15\% = 85\%$

$\dfrac{85}{100} = 0.85$

$75.65 \div 0.85 = 89$

The original price was £89

EXAM ALERT!

Do **not** increase the price given by 15%.

Subtract to find the multiplier for a 15% decrease then **divide** the new amount by the multiplier.

Check it!

Reduce £89 by 15%:

$£89 \times 0.85 = £75.65$ ✓

Students have struggled with exam questions similar to this – **be prepared!**

Compound interest

If you leave your money in a bank account it will earn compound interest.

This table shows the total interest earned after 3 years at 3% compound interest.

Year	Balance (£)	Interest earned (£)
1	40 000	1200
2	41 200	1236
3	42 436	1273.08
	43 709.08	3709.08

For year 2 you have to calculate 3% of £41 200

After 3 years the total interest earned will be £3709.08 and the balance will be £43 709.08

Repeated change

Compound interest is an example of REPEATED PERCENTAGE CHANGE. You can use multipliers to calculate repeated percentage changes quickly.

This car DEPRECIATES in value by 8% each year.

£15000

The multiplier for an 8% decrease is 0.92

After 3 years the car is worth:

$£15\,000 \times 0.92 \times 0.92 \times 0.92 = £11\,680.32$

Now try this

The multiplier for a 2% increase is 1.02

1 Hannah buys some shoes in a sale where all the items are marked '40% off'. She pays £27 for the shoes. What price were the shoes originally?

(3 marks)

target B

2 The population of a city is 183 000. The population is increasing at the rate of 2% per year. Estimate the population in 5 years' time. Give your answer to 3 significant figures.   *(3 marks)*

target B

A*
A
B
C
D

Ratio

RATIOS are used to compare quantities.

You can find equivalent ratios by multiplying or dividing by the same number.

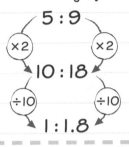

5 : 9

×2 ... ×2

10 : 18

÷10 ... ÷10

1 : 1.8

This equivalent ratio is in the form 1 : *n*. This is useful for calculations.

Simplest form

To write a ratio in its simplest form, find an equivalent ratio with the smallest possible whole number values.

Simplest form
5 : 1   10 : 9
2 : 3 : 4

NOT simplest form
10 : 2   1 : 0.9
1 : 1.5 : 2

Worked example

target C

Alexis and Nisha share a flat. They decide to split their phone bill in the ratio 3 : 5.

Alexis pays £78. How much does Nisha pay?   *(3 marks)*

78 ÷ 3 = 26

26 × 5 = 130

Nisha pays £130

Work out the value of one part of the ratio first. You could also use equivalent ratios to solve this problem. Work out how many 3s go into 78 to find the multiplier.

3 : 5

×26 ... ×26

78 : 130

Check it!

£130 + £78 = £208

Divide £208 in the ratio 3 : 5

3 + 5 = 8 parts in the ratio in total

£208 ÷ 8 = £26

£26 × 3 = £78 ✓

Worked example

target C

Karl is comparing two juice drinks:

JAFFA JUICE
25% Fruit Juice

SUPER SLURP
Ratio of fruit juice to water
5:13

Which drink contains a greater proportion of fruit juice?   *(2 marks)*

$\frac{5}{18} \times 100 = 27.777\ldots$

27.8% of Super Slurp is fruit juice. So Super Slurp has a greater proportion of fruit juice.

Convert 5 : 13 into a fraction. There are 5 + 13 = 18 parts in the ratio so the denominator of the fraction is 18. Convert your fraction into a percentage and compare the two proportions.

Now try this

target C

1 A recipe for scones uses self-raising flour and butter in the ratio 15 : 4.

Delia uses 420 g of self-raising flour.

How much butter should she use?   *(3 marks)*

Use equivalent ratios and work out how many 15s go into 420.

target C

2 A lemon and lime drink is made from lemonade and lime juice in the ratio 13 : 3. What percentage of the drink is lime juice?   *(3 marks)*

Standard form

A*
A
B
C
D

Numbers in standard form have two parts.

$$7.3 \times 10^{-6}$$

This part is a number greater than or equal to 1 and less than 10

This part is a power of 10

You can use standard form to write very large or very small numbers.

$$920\,000 = 9.2 \times 10^5$$

Numbers greater than 10 have a positive power of 10

$$0.007\,03 = 7.03 \times 10^{-3}$$

Numbers less than 1 have a negative power of 10

Counting decimal places

You can count decimal places to convert between numbers in standard form and ordinary numbers.

3 jumps

$$7\,9\,0\,0 = 7.9 \times 10^3$$

| 7900 > 10 |
| So the power |
| is positive |

4 jumps

$$0.0\,0\,0\,3\,5 = 3.5 \times 10^{-4}$$

| 0.00035 < 1 |
| So the power |
| is negative |

BE CAREFUL!
Don't just count zeros to work out the power.

Worked example

(a) Write 1 630 000 in standard form.   *(1 mark)*

1.63×10^6

(b) Write 4.2×10^{-3} as an ordinary number.   *(1 mark)*

0.0042

Count the number of decimal places you need to move to get a number between 1 and 10. 1 630 000 is bigger than 10 so the power will be positive.

Using a calculator

You can enter numbers in standard form using the ×10ˣ key.

To enter 3.7×10^{-6} press

[3] [.] [7] [×10ˣ] [(−)] [6]

If you are using a calculator with numbers in standard form it is a good idea to put brackets around each number.

For part (b) you would enter:

[(] [1] [.] [9] [×10ˣ] [4] [)] [x²] [=]

Worked example

A and B are standard form numbers.
$A = 1.9 \times 10^4$ $B = 4.2 \times 10^5$

Calculate, giving your answers in standard form:

(a) $A + B$ *(1 mark)*

$(1.9 \times 10^4) + (4.2 \times 10^5) = 439\,000$
$$= 4.39 \times 10^5$$

(b) A^2 *(1 mark)*

$(1.9 \times 10^4)^2 = 3.61 \times 10^8$

Now try this

1 (a) Write 0.00028 in standard form.
   (1 mark)

(b) Write 3.91×10^5 as an ordinary number.   *(1 mark)*

$\dfrac{pq}{5}$ means $(p \times q) \div 5$

2 p and q are standard form numbers.

$p = 3.5 \times 10^6$ $q = 7.9 \times 10^5$

Calculate, giving your answers in standard form:

(a) $p - q$ *(1 mark)*

(b) $\dfrac{pq}{5}$ *(1 mark)*

Upper and lower bounds

Upper and lower bounds are a measure of accuracy. For example, the width of a postcard is given as 8 cm to the nearest cm.

lower bound   upper bound

7 cm   7.5 cm   8 cm   8.5 cm   9 cm

The actual width of the postcard could be anything between 7.5 cm and 8.5 cm.

7.5 cm is called the LOWER BOUND or MINIMUM VALUE the width could be.

8.5 cm is called the UPPER BOUND or MAXIMUM VALUE the width could be.

Using upper and lower bounds in calculations

To find the greatest possible and least possible values of a calculation use these rules.

	+	−	×	÷
Greatest value	UB + UB	UB − LB	UB × UB	UB ÷ LB
Least value	LB + LB	LB − UB	LB × LB	LB ÷ UB

Least value of $a + b$ = lower bound of a + lower bound of b

Worked example

Charlie has a piece of ribbon measuring 100 cm to two significant figures. He cuts a piece off the end of the ribbon measuring 21 cm to the nearest cm.

Charlie needs 70 cm of ribbon for another project. Will he definitely have enough left over?   *(3 marks)*

	Lower bound	Upper bound
Length of ribbon	95 cm	105 cm
Length of piece cut off	20.5 cm	21.5 cm

Minimum length remaining = 95 − 21.5
$$= 73.5 \text{ cm}$$

Charlie will definitely have enough ribbon for his next project.

Degree of accuracy

When tackling the most demanding questions, you might need to give a value to 'an appropriate degree of accuracy'.

upper bound of x = 1.17956

lower bound of x = 1.17892

2 decimal places would be an appropriate degree of accuracy for x. The upper and lower bounds both round to 1.18 to 2 decimal places.

Now try this

1 Jenson drives 280 miles, to the nearest 10 miles, in 4 hours, to the nearest hour.

Work out his maximum average speed in miles per hour.   *(3 marks)*

2 A cylinder has a volume of 115 cm³, to the nearest cm³. It's radius is 2.3 cm, correct to 1 decimal place.

Work out the minimum height of the cylinder.

Give your answer to an appropriate degree of accuracy.   *(4 marks)*

The Data Handling Cycle

A*
A
B
C
D

In your Unit 1 exam, you might have to write a plan for a statistical investigation.
A statistical investigation always follows the four components of the DATA HANDLING CYCLE.

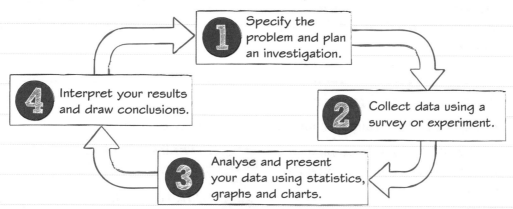

1 Specify the problem and plan an investigation.

2 Collect data using a survey or experiment.

3 Analyse and present your data using statistics, graphs and charts.

4 Interpret your results and draw conclusions.

Hypothesis testing

In statistics, a hypothesis is a statement that might be either true or false. You can TEST whether the hypothesis is true by carrying out a statistical investigation.

Golden rule

When you're answering questions using the Data Handling Cycle, make sure your answers are specific to the hypothesis you want to test.

Worked example

target **D**

Sean wants to test this hypothesis:

> Boys have larger handspans than girls.

Use the Data Handling Cycle to write a plan for Sean.   *(3 marks)*

Collect data by measuring the handspans of 15 boys and 15 girls in my class to the nearest mm.

Calculate the mean and median handspan for the girls and the mean and median handspan for the boys.

Compare the averages for the boys and the girls and write a conclusion.

Use the four components of the Data Handling Cycle:

1 The problem has been given in the question so you don't need to write anything for this component.

2 Write down one way you could collect data such as a survey or an experiment.

3 Write down at least one way you could analyse or present your data, such as calculating statistics or drawing graphs.

4 Describe how you will use your statistics or graphs to interpret your results.

Now try this

target **D**

Sam wants to test this hypothesis:

> Girls use social networking sites more often than boys.

Use the Data Handling Cycle to suggest how she should test this hypothesis.   *(3 marks)*

Collecting data

You can collect PRIMARY data yourself, or get SECONDARY data from another source like the Internet. Data which is recorded in words, like the make of a car, is called QUALITATIVE data. Data which is recorded in numbers is called QUANTITATIVE data. Quantitative data can be DISCRETE, like the number of CDs you own, or CONTINUOUS, like time or length.

Methods of data collection

Here are examples of the main four methods of data collection:

1 OBSERVATION: Watching what foods students buy most often in the canteen.

2 EXPERIMENT: Measuring skin temperature before and after exercise.

3 SURVEY or QUESTIONNAIRE: Asking local people questions about their shopping habits.

4 DATA LOGGING: Recording the number of cars that use a section of road.

Worked example

target D

These two frequency tables show the same data.

A
Shoe size	7	$7\frac{1}{2}$	8	$8\frac{1}{2}$	9	$9\frac{1}{2}$	10	$10\frac{1}{2}$	11	$11\frac{1}{2}$	12
Frequency	2	4	5	4	11	5	7	7	4	3	1

B
Shoe size	$7-8\frac{1}{2}$	$9-10\frac{1}{2}$	$11-12\frac{1}{2}$
Frequency	15	30	8

> You could also say that you can calculate the exact mean from table A, whereas you can only calculate an estimate of the mean from table B.

(a) Give **one** advantage of table A over table B.   *(1 mark)*

Table A shows the exact data which was collected.

(b) Give **one** advantage of table B over table A.   *(1 mark)*

Table B shows a summary of the data which is easier to read.

Now try this

target D

State whether the following investigations involve primary or secondary data and also whether the data is discrete or continuous.

Put a tick in the appropriate places in the table.

Investigation	Primary	Secondary	Discrete	Continuous
(a) In Wales in 2011, more people bought new cars in March than in September.				
(b) Carlisle has more rainfall in October than Swansea.				
(c) More people in my year group have a dog as a pet than a cat.				

(3 marks)

8

Surveys

You need to know what makes a question good or bad in a survey.
Look at this example then read the comments.

Internet use survey

1. What do you use to access the internet?

   ....

2. How much time do you spend on the internet each day?

   ☐ Not very much   ☐ Average   ☐ A lot

3. How many times a week do you check your email?

   ☐ 1–5   ☐ 5–10   ☐ 10–15   ☐ Every day

4. Have you ever downloaded films illegally from the internet?

   ....

5. Do you agree that the BBC iPlayer is very easy to use?

   ☐ Yes   ☐ No

✗ It's hard to know what this question means. Add at least 4 response boxes to improve the question:
☐ Laptop computer
☐ Desktop computer
☐ Smartphone
☐ Other

✗ People aren't very likely to answer this question truthfully. Don't ask people to reveal embarrassing or personal information.

✗ These responses could mean different things to different people. It would be better to ask how many hours they spend on the internet each day.

✗ The response boxes overlap. These response boxes would be better:
☐ 0–5   ☐ 6–10
☐ 11–15
☐ 16 or more

✗ This is a **biased** question. People are more likely to agree with you. A better question would be: The BBC iPlayer is easy to use.
☐ Agree   ☐ Disagree
☐ Neither

Worked example

Every Friday, a company offers free breakfasts to employees who cycle to work. The managing director thinks that 50% more people cycle to work on a Friday than on any other day.

Design an observation sheet the managing director can use to see if she is right. *(2 marks)*

Number of people who cycle

Day of the week	Tally	Frequency
Monday		
Tuesday		
Wednesday		
Thursday		
Friday		

EXAM ALERT!

An observation sheet is sometimes called a **data collection sheet**. Imagine what information you would need to collect to investigate the hypothesis. You need to refer to the **context** of the question. You can do this by writing a heading explaining that your tally chart shows the number of people who cycle.

Students have struggled with exam questions similar to this – **be prepared!**

Now try this

Helen thinks more people in her school buy lunch rather than bringing it from home.

(a) Design an observation sheet Helen can use to see if she is right. *(2 marks)*

(b) Helen wants to carry out a survey to find out how much students spend on lunch.

Write down a question she could use.

Include a response section. *(2 marks)*

Two-way tables

Two-way tables are easy marks in your exam. You can answer these questions by adding or subtracting.

	Year 7	Year 8	Year 9	Total
Vegetarian	14	22	25	61
Not vegetarian	72	63	54	189
Total	86	85	79	250

There were 61 vegetarians in total.

In total 250 students were surveyed.

There were 86 Year 7 students surveyed.

There were 63 non-vegetarians in Year 8.

Worked example

D

Anton surveyed 120 people about how they voted at the last general election. He recorded the results in a two-way table:

	Labour	Conservative	Other	Total
Female	21	13	13	47
Male	32	27	14	73
Total	53	40	27	120

Complete the two-way table. *(4 marks)*

Labour column: 53 − 21 = 32
Female row: 47 − 21 − 13 = 13
Conservative column: 13 + 27 = 40
Total row: 120 − 53 − 40 = 27
Other column: 27 − 13 = 14
Male row: 32 + 27 + 14 = 73
Check: 47 + 73 = 120
53 + 40 + 27 = 120 ✓

Everything in red is part of the answer.

Golden rules

The numbers in each column add up to the total for that column.

Other	
	13
	+ 14
	= 27

The numbers in each row add up to the total for that row.

Female	21	+ 13	+ 13	= 47

To complete a two-way table:
• write the total in the bottom right-hand cell
• look for rows and columns with only one missing number
• use addition and subtraction to find any missing values
• fill in the missing values as you go along.

Check it!
Add up the row totals and the column totals. They should be the same.

Now try this

C

Here is some information about class 9H.

There are 32 students altogether.

There are 4 more girls than boys

A third of the girls are left-handed.

There are 23 right-handed students altogether.

Copy and complete the two-way table to show this information.

	Boys	**Girls**	**Total**
Left-handed			
Right-handed			
Total			32

(3 marks)

A* A A B C D

Sampling

Sample

A SAMPLE is a small group chosen from a larger population.

The red figures represent a sample of 4 from a population of 16.

You can make conclusions about a population by collecting data from a sample.

It is usually cheaper and quicker to collect data from a sample.

Random sample

One way of reducing BIAS in a statistical investigation is to use a RANDOM SAMPLE. In a random sample every member of the population has an equal chance of being selected. You could select a random sample by numbering the population and selecting numbers at random using a computer or a random number table.

Stratified sampling

A stratified sample is one in which the population is split into groups. A simple random sample is taken from each group. The number taken from each group should be in proportion to the size of the group.

There are twice as many boys as girls in this population...

... so you need twice as many boys as girls in a stratified sample.

Girls   Boys

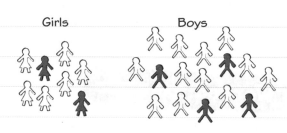

Worked example

 target A

A tennis club has 255 male members and 346 female members. Alison is carrying out a survey of these members. She uses a sample of 50 members, <u>stratified</u> by gender.

Work out the number of female members Alison should select for her sample. *(3 marks)*

$$255 + 346 = 601$$

$$\frac{346}{601} \times 50 = 28.7853\ldots$$

Alison should select 29 female members.

EXAM ALERT!

If you see the word 'stratified' in a question underline it.

You could also work this out by finding the **sampling fraction**. Divide the sample size by the total population to work out the sampling fraction, then multiply by the number of female members in the club.

$$\frac{50}{601} \times 346 = 28.7853\ldots$$

Remember to give your final answer as a whole number.

Students have struggled with exam questions similar to this – **be prepared!**

Now try this

 target A

A golf club has 720 members.

A stratified sample of members is taken by age group.

Complete the table opposite.

Age group	Junior	18–39	40–59	Senior
Number of members	120			150
Number in sample	20		45	

(3 marks)

A*
A
B
C
D

Mean, median and mode

You can analyse data by calculating statistics like the MEAN, MEDIAN and MODE.

Mean

Add up all the values

↓

Divide by the total number of values

↓

Do not round your answer

Median

Write the values in order of size, smallest first

↓

Count the number of values

Odd number of values →
The median is the middle value

Even number of values →
The median is half-way between the two middle values

Mode

Look for the most common value

One value with highest frequency →
This value is the mode

More than one value with highest frequency →
These values are all modes

All values have same frequency →
There is no mode

Worked example

 target C

Kayla has 8 numbered cards.

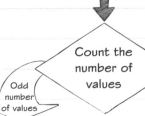

She removes two cards. The mean value of the remaining cards is 4.

Which two cards could Kayla have removed? Give **one** possible answer.
(4 marks)

$6 \times 4 = 24$

$1 + 2 + 3 + 4 + 5 + 6 + 7 + 8 = 36$

$36 - 24 = 12$

The removed cards add up to 12
So Kayla could have removed 7 and 5

Check:

$$\frac{1 + 2 + 3 + 4 + 6 + 8}{6} = 4 ✓$$

You can work out the sum of the 6 remaining cards using this formula:

Sum of values = mean × number of values

Subtract this sum from the sum of all 8 cards. This tells you the sum of the 2 cards Kayla removed. The removed cards were either 5 and 7 or 8 and 4.

Check it!
Work out the mean of the remaining 6 cards.

Which average works best?

	👍	👎
Mean	Uses all the data	Affected by extreme values
Median	Not affected by extreme values	Value may not exist
Mode	Suitable for data that can be described in words	Not always near the middle of the data

---

Now try this

 target C

Joe scored these marks in **six** Maths tests.

11   9   5   13   15   12

How many marks must he score in the next test so that his mean mark and his median mark are the same?
(3 marks)

Make sure you check your answer by calculating the new mean and median.

A*
A
B
C
D

Frequency table averages

Finding averages from frequency tables and frequency polygons is a common exam question.
This frequency table shows the numbers of pets owned by the students in a class.

The mode is 1.
This value has
the highest
frequency.

Number of pets (x)	Frequency (f)	Frequency × number of pets (f × x)
O	12	12 × O = O
1	18	18 × 1 = 18
2	5	5 × 2 = 10
3	2	2 × 3 = 6
Total	37	34

To calculate
the mean you
need to add
a column for
'f × x'.

There are 37 values so the median is
the $\frac{37+1}{2} = 19$th value.

The first 12 values are all O. The next 18
values are 1. So the median is 1.

The total in the 'f × x' column represents the total number
of pets owned by the class.

$$\text{Mean} = \frac{\text{total number of pets}}{\text{total frequency}} = \frac{34}{37} = 0.92 \text{ (to 2 d.p.)}$$

Worked example

Maisie recorded the times, in minutes, taken by 150 students to travel to school.
The table shows her results.

Time (*t* minutes)	Frequency (f)	Midpoint (x)	f × x
0 ≤ t < 20	65	10	65 × 10 = 650
20 ≤ t < 30	40	25	40 × 25 = 1000
30 ≤ t < 40	39	35	39 × 35 = 1365
40 ≤ t < 60	6	50	6 × 50 = 300

Total frequency = 150   Total of f × x = 3315

(a) Work out an estimate for the mean number of
minutes that the students took to travel to school.
(4 marks)

$$\frac{3315}{150} = 22.1$$

(b) Explain why your answer to part **(a)** is an estimate.
(1 mark)

Because you don't know the exact data values.

Everything in red is
part of the answer.

EXAM ALERT!

Add extra columns to the table for
'Midpoint (x)' and
'Midpoint × frequency (f × x)'.

$$\text{Estimate of mean} = \frac{\text{Total of } fx \text{ column}}{\text{Total frequency}}$$

Students have struggled with
exam questions similar to
this – **be prepared!**

Now try this

The table shows the ages of 120 people in a
small village.

Work out an estimate of the mean age of these
120 people.   *(4 marks)*

Age, x (years)	Frequency
0 < x ≤ 20	33
20 < x ≤ 40	27
40 < x ≤ 70	39
70 < x ≤ 100	21

A*
A
B
C
D

Interquartile range

Range and interquartile range are measures of spread. They tell you how spread out data is.

QUARTILES divide a data set into four equal parts.

Half of the values lie between the lower quartile and the upper quartile.

|← Interquartile range (IQR) →|

$Q_1 = \dfrac{n+1}{4}$th value,

where n = number of data values

× ×   × × × ×   × × × ×   × DATA VALUES

Smallest value   Lower quartile (Q_1)   Median (Q_2)   Upper quartile (Q_3)   Largest value

$Q_3 = \dfrac{3(n+1)}{4}$th value

RANGE = largest value − smallest value

INTERQUARTILE RANGE (IQR) = upper quartile (Q_3) − lower quartile (Q_1)

Worked example

target B

Alison recorded the heights, in cm, of some tree saplings. She put the heights in order.

21   23   23   25   26   26   31   32
33   35   36   40   40   41   42

Work out the interquartile range of Alison's data.

$n = 15$

$\dfrac{n+1}{4} = \dfrac{15+1}{4} = 4$

$Q_1 = $ 4th value $= 25\,cm$

$\dfrac{3(n+1)}{4} = \dfrac{3(15+1)}{4} = 12$

$Q_3 = $ 12th value $= 40\,cm$

$IQR = Q_3 - Q_1 = 40 - 25 = 15\,cm$

To work out the interquartile range, you need to know the lower quartile and the upper quartile.

1. Count the total number of values, n.

2. Check that the data is arranged in order of size.

3. Find the $\dfrac{n+1}{4}$th data value. This is the lower quartile (Q_1).

4. Find the $\dfrac{3(n+1)}{4}$th data value. This is the upper quartile (Q_3).

5. Subtract the lower quartile from the upper quartile to find the interquartile range.

Golden rule

Always arrange the data in order of size before calculating the median or quartiles.

If the data is given in an ordered STEM-AND-LEAF DIAGRAM then it is already in order of size. This stem-and-leaf diagram shows the costs, in £, of some DVDs. There are 11 pieces of data in this stem-and-leaf diagram.

0 | 9 represents £9

This is the STEM.

```
0 | 7  9  9
1 | 0  0  2  3  5  7
2 | 0  5
```

Key: 1 | 5 = £15

There are 11 pieces of data, so the median is the 6th value.
The median is £12.

Now try this

target D-B

The number of driving lessons taken by 27 people before passing their driving test, are shown.

23   14   10   16   34   21   19   8   14
16   31   20   23   45   28   18   19   20
15   12   26   12   40   32   36   22   16

(a) Complete an ordered stem and leaf diagram to represent this data. Remember to include a key. *(3 marks)*

(b) Work out the median and the interquartile range for this data. *(3 marks)*

A*
A
B
C
D

Scatter graphs

The points on a scatter graph aren't always scattered. If the points are almost on a straight line then the scatter graph shows CORRELATION. The better the straight line, the stronger the correlation.

Negative correlation

No correlation

Positive correlation

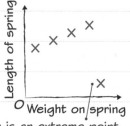

An isolated point on a scatter graph is an extreme point that lies outside the normal range of values.

Worked example

target
D

This scatter graph shows the engine capacity of some cars and the distance they will travel on one gallon of petrol.

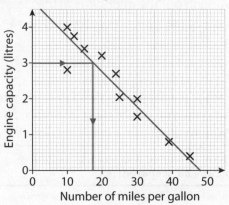

(a) Isaac has the hypothesis:

> Cars with smaller engines are more efficient.

Comment on his hypothesis.   *(1 mark)*

The graph shows negative correlation so there is strong support for his hypothesis.

(b) Estimate how far a car with a 3 litre engine will travel on 1 gallon of petrol.   *(2 marks)*

17 miles

(a) There is negative correlation, so cars with bigger engines travel less far on one gallon of petrol. This supports Isaac's hypothesis. You can use words like 'strong' and 'weak' to show how much support the data provides for the hypothesis.

(b) Draw a line of best fit on the scatter graph. Read across from 3 on the vertical axis to the line of best fit, then down to the horizontal axis. Draw all the lines you use on the graph.

Line of best fit checklist

Straight line that is as close as possible to all the points.   ✓

Used to predict values.   ✓

Does not need to go through (0, 0).   ✓

Drawn with a ruler.   ✓

Ignores isolated points.   ✓

Now try this

target
D

This scatter graph shows the daily hours of sunshine and the daily maximum temperature at 12 seaside resorts in England on one day last summer.

(a) Draw a line of best fit for this data.   *(1 mark)*

(b) Is the correlation positive or negative?   *(1 mark)*

(c) Describe the connection between the number of hours of sunshine and the maximum temperature.   *(1 mark)*

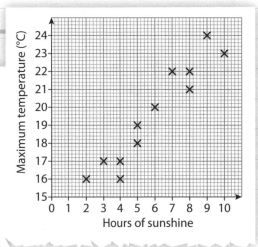

A*
A
B
C
D

Frequency polygons

You can represent grouped data using a FREQUENCY POLYGON. Look at this example.

Reaction time (r milliseconds)	Frequency
$100 \leqslant r < 200$	7
$200 \leqslant r < 300$	15
$300 \leqslant r < 400$	10

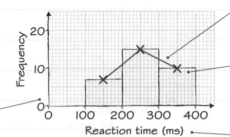

Join the points with STRAIGHT LINES. Make sure you use a ruler.

Plot points at the MIDPOINT of each class interval.

You always record FREQUENCY on the vertical axis.

If you draw a histogram on the same graph the frequency polygon joins together the midpoints of the tops of the bars.

This frequency polygon shows the reaction times of a class of students.

Worked example

C (target)

30 students timed how long it took them to complete a jigsaw puzzle. The results were recorded in a grouped frequency table:

Time (t minutes)	Frequency	Midpoint
$10 \leqslant t < 14$	2	12
$14 \leqslant t < 18$	5	16
$18 \leqslant t < 22$	12	20
$22 \leqslant t < 26$	8	24
$26 \leqslant t < 30$	3	28

Show this information on a frequency polygon. *(3 marks)*

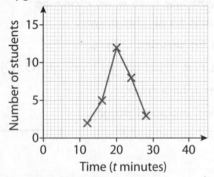

EXAM ALERT!

Start by working out the midpoints of the class intervals. The midpoint of the class interval $10 \leqslant t < 14$ is $\dfrac{10 + 14}{2} = 12$

Check it!
In your exam you will only be asked to draw a frequency polygon for data with **equal class intervals**. So make sure that your midpoints are the same distance apart.

Students have struggled with exam questions similar to this – **be prepared!**

Estimating the mean

You can estimate the mean for data given in a frequency polygon using this formula:

$$\text{Mean} \approx \frac{\text{Sum of (midpoint} \times \text{frequency)}}{\text{Total frequency}}$$

For the worked example on the left:

$$\frac{(2 \times 12) + (5 \times 16) + (12 \times 20) + (8 \times 24) + (3 \times 28)}{30} = 20.6666...$$

An estimate for the mean time is 20 minutes, 40 seconds.

There is more about estimating the mean of grouped data on page 13.

Now try this

C-B (target)

This table shows the times taken, in minutes, for 50 people to solve a crossword puzzle.

(a) Draw a frequency polygon for this data. *(2 marks)*

(b) Which interval contains the median time? *(1 mark)*

Time, t (minutes)	Frequency
$0 < t \leqslant 10$	3
$10 < t \leqslant 20$	9
$20 < t \leqslant 30$	11
$30 < t \leqslant 40$	18
$40 < t \leqslant 50$	7
$50 < t \leqslant 60$	2

A*
A
B
C
D

Histograms

Histograms are a good way to represent grouped data with DIFFERENT class widths.

Worked example

This table shows the finishing times in minutes of runners in a cross-country race.

Time (t minutes)	Frequency	Frequency density
$16 \leqslant t < 20$	12	3
$20 \leqslant t < 30$	45	4.5
$30 \leqslant t < 50$	28	1.4

Draw a suitable graph to represent the data.

(3 marks)

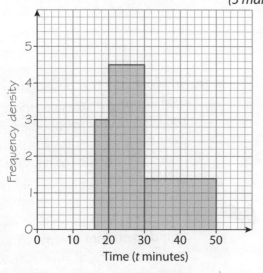

Histogram facts

No gaps between the bars. ✓

AREA of each bar is proportional to frequency. ✓

Vertical axis is labelled 'Frequency density'. ✓

Bars can be different widths. ✓

Frequency density = $\dfrac{\text{frequency}}{\text{class width}}$ ✓

The class widths in the frequency table are **different widths** so a histogram is the most suitable graph to draw.

1. Calculate the frequency densities.

$$\frac{12}{4} = 3 \qquad \frac{45}{10} = 4.5 \qquad \frac{28}{20} = 1.4$$

2. Add these values to the table as an extra column.

3. Label the vertical axis 'Frequency density'.

4. Choose a scale for the vertical axis and draw each bar.

Area and estimation

You can use the area under a histogram to estimate frequencies. An estimate for the NUMBER of maggots between 1 mm and 2 mm long is:

$$0.5 \times 22 + 0.5 \times 6 = 14$$

You might need to answer proportion questions about histograms in your exam. The total frequency is:

$$1 \times 14 + 0.5 \times 22 + 1.5 \times 6 = 34$$

So an estimate for the PROPORTION of maggots between 1 mm and 2 mm long is: $\dfrac{14}{34} = 0.4117\ldots$ or 41% (2 s.f.)

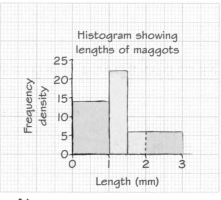

Now try this

A speed camera recorded the speed of some vehicles on a motorway. The table on the right shows the results.

(a) On graph paper, draw a histogram to illustrate this data. *(3 marks)*

(b) Estimate the proportion of vehicles travelling at more than 65 mph. *(2 marks)*

Speed, s (mph)	Frequency
$0 < s \leqslant 40$	48
$40 < s \leqslant 50$	32
$50 < s \leqslant 70$	128
$70 < s \leqslant 80$	88
$80 < s \leqslant 110$	24

A* A B C D

Probability 1

For EQUALLY LIKELY OUTCOMES the probability (P) that something will happen is:

$$\text{Probability} = \frac{\text{number of successful outcomes}}{\text{total number of possible outcomes}}$$

If you know the probability that an event WILL happen, you can calculate the probability that it won't happen:

P(Event doesn't happen) = 1 − P(Event happens)

The probability of rolling a 6 on a normal fair dice is $\frac{1}{6}$. So the probability of NOT rolling a 6 is $1 - \frac{1}{6} = \frac{5}{6}$

Add or multiply?

Events are MUTUALLY EXCLUSIVE if they can't BOTH happen at the same time. For mutually exclusive events:

P(A or B) = P(A) + P(B)

Events are INDEPENDENT if the outcome of one doesn't affect the outcome of the other. For independent events:

P(A and B) = P(A) × P(B)

Worked example

target D

Amir designs a game for his school fete. This table shows the probability of winning a prize:

Prize	Badge	Keyring	Cuddly toy
Probability	0.35	0.18	0.07

(a) What is the probability of **not** winning a prize? *(1 mark)*

P(Win) = 0.35 + 0.18 + 0.07 = 0.6
P(Not win) = 1 − 0.6 = 0.4

(b) Amir plays the game three times. What is the probability that he does not win a prize? *(2 marks)*

0.4 × 0.4 × 0.4 = 0.064

(a) To work out the probability of winning **any** prize you need to add together the probabilities. Then you can use this rule to work out the probability of not winning a prize:
$$P\binom{\text{NOT winning}}{\text{a prize}} = 1 - P\binom{\text{Winning a}}{\text{prize}}$$

(b) Each time Amir plays the game is an **independent** event. To work out the probability of Amir not winning three times you need to multiply the probabilities.

Sample space diagrams

First coin

	H	T
H	HH	TH
T	HT	TT

Second coin

A SAMPLE SPACE DIAGRAM shows you all the possible outcomes of an event. Here are all the possible outcomes when two coins are flipped.

• There are four possible outcomes. TH means getting a tail on the first coin and a head on the second coin.

• The probability of getting two tails when two coins are flipped is $\frac{1}{4}$ or 0.25. There are 4 possible outcomes and only 1 successful outcome (TT).

Now try this

target D-C

A spinner has four colours. This table shows the probability of landing on each colour.

Colour	Red	Blue	Green	Yellow
Probability	0.1		0.2	0.1

(a) What is the probability of landing on blue? *(1 mark)*

(b) What is the probability of not landing on red? *(1 mark)*

(c) The spinner is spun twice. What is the probability that it lands on blue both times? *(2 marks)*

Each spin is an independent event so you can multiply the probabilities.

Probability 2

You need to be able to calculate probabilities for data given in graphs and tables. You can use this formula to estimate a probability from a frequency table:

$$\text{Probability} = \frac{\text{frequency of outcome}}{\text{total frequency}}$$

This is called RELATIVE FREQUENCY.

Golden rule

Probability estimates based on relative frequency are MORE ACCURATE for larger samples (or for more trials in an experiment).

In the sample there were $15 + 10 = 25$ eggs which weighed 55 g or more. So an estimate for the probability of picking an egg which weighs 55 g or more is $\frac{25}{40}$ or $\frac{5}{8}$.

When you are calculating probabilities you can give your answers as decimals or fractions in lowest terms.

Worked example

An egg farm weighed a sample of 40 eggs. It recorded the results in a frequency table:

Weight, w (g)	Frequency
$45 \leqslant w < 50$	6
$50 \leqslant w < 55$	9
$55 \leqslant w < 60$	15
$60 \leqslant w < 65$	10

(a) Roselle buys some eggs from the farm and picks one at random. Estimate the probability that the egg weighs 55 g or more. *(2 marks)*

$15 + 10 = 25$

$P(w \geqslant 55) \approx \dfrac{25}{40} = \dfrac{5}{8}$

So the probability is $\dfrac{5}{8}$

(b) Comment on the accuracy of your estimate. *(1 mark)*

40 is a fairly small sample size, so the estimate is not very accurate.

Expectation

Probability helps you predict the outcome of an event.

If you flip a coin 100 times, you can expect to get heads about 50 times. You probably won't get heads exactly 50 times, but it's a good guess.

$$\text{Expected number of outcomes} = \text{number of trials} \times \text{probability}$$

Worked example

The probability of a biased coin landing heads up is 0.4

The coin is flipped 300 times.

Work out an estimate for the number of times the coin will land heads up.

$300 \times 0.4 = 120$

Now try this

In a school, the probability that a girl is left-handed is 0.15
The probability that a boy is left-handed is 0.2
The school has 420 girls and 390 boys.

An estimate for the number of left-handed **girls** is 420×0.15

(a) Estimate the number of left-handed students in the school. *(4 marks)*

(b) A student is picked at random from the whole school. Estimate the probability that the student is left-handed. *(1 mark)*

Tree diagrams

A tree diagram shows all the possible outcomes from a series of events and their probabilities.

This is a tree diagram for Holly's journey to school.

You write the probability for each event on the branch.

At each branch the probabilities add up to 1.
$\frac{2}{3} + \frac{1}{3} = 1$

The outcome of the first event can affect the probability of the second.

Holly is less likely to be on time if she misses the bus.

Each branch is like a different parallel universe. In this universe, Holly catches the bus and gets to school on time.

You write the outcomes at the ends of the branches.
You can use shorthand like this.

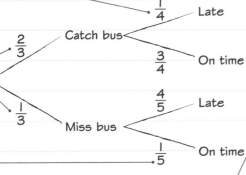

Outcome	Probability
CL	$\frac{2}{3} \times \frac{1}{4} = \frac{2}{12} = \frac{1}{6}$
CO	$\frac{2}{3} \times \frac{3}{4} = \frac{6}{12} = \frac{1}{2}$
ML	$\frac{1}{3} \times \frac{4}{5} = \frac{4}{15}$
MO	$\frac{1}{3} \times \frac{1}{5} = \frac{1}{15}$

You multiply along the branches to find the probability of each outcome.

The probability that Holly misses the bus and is late for school is $\frac{4}{15}$.

Golden rules

1 Look out for the words REPLACE or PUT BACK in a probability question.

WITH replacement: probabilities stay the same.

WITHOUT replacement: first probability stays the same while the others change.

2  MULTIPLY ALONG THE BRANCHES ADD UP THE OUTCOMES

Worked example

 target A*

There are 3 strawberry yoghurts and 4 pineapple yoghurts in a fridge. Noah picks two yoghurts at random from the fridge. Work out the probability that both the yoghurts were the same flavour. *(4 marks)*

This is an example of selection **without replacement**. The two events are not independent. The probabilities for the second pick change depending on which flavour yoghurt was picked first. A tree diagram is the **safest** way to answer questions like this.

First yoghurt　Second yoghurt　Outcome　Probability

$\frac{3}{7}$ S

$\frac{2}{6}$ S　　SS　　$\frac{3}{7} \times \frac{2}{6} = \frac{1}{7}$

$\frac{4}{6}$ P　　SP　　$\frac{3}{7} \times \frac{4}{6} = \frac{2}{7}$

$\frac{4}{7}$ P

$\frac{3}{6}$ S　　PS　　$\frac{4}{7} \times \frac{3}{6} = \frac{2}{7}$

$\frac{3}{6}$ P　　PP　　$\frac{4}{7} \times \frac{3}{6} = \frac{2}{7}$

P (both yoghurts same flavour) = P(SS) + P(PP)

$= \frac{1}{7} + \frac{2}{7} = \frac{3}{7}$

Now try this

target A*

Grace buys a packet of 12 tulip bulbs. They all look the same, but 7 of them will produce red flowers and 5 will produce yellow flowers.

A bulb is taken at random and planted.

A second bulb is taken at random and planted. Work out the probability that the two bulbs produce **different** coloured flowers. *(4 marks)*

Cumulative frequency

In your exam you might have to draw a cumulative frequency graph, or use one to find the median or the interquartile range.

How to draw a cumulative frequency graph

Reaction time t (s)	Frequency	Cumulative frequency
$0 < t \leqslant 0.1$	2	2
$0.1 < t \leqslant 0.2$	5	$2 + 5 = 7$
$0.2 < t \leqslant 0.3$	18	$7 + 18 = 25$
$0.3 < t \leqslant 0.4$	5	$25 + 5 = 30$
$0.4 < t \leqslant 0.5$	1	$30 + 1 = 31$

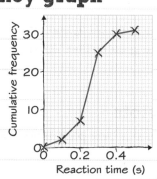

1. Plot 0 at the beginning of the first class interval.

2. Plot each value at the UPPER end of its class interval.

3. Join your points with a ruler or a smooth curve.

Add a column for CUMULATIVE FREQUENCY to your frequency table.

Check that your final value is the same as the total frequency.

On your exam paper, cumulative frequency graphs will usually be drawn with straight lines.

Here's another example:

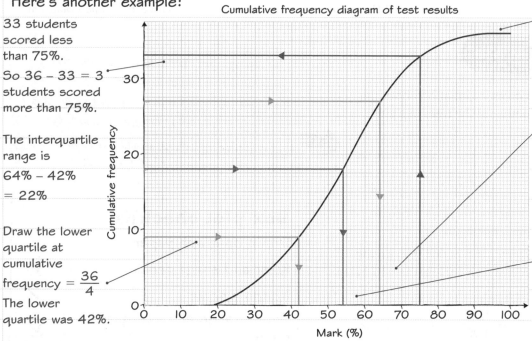

Cumulative frequency diagram of test results

33 students scored less than 75%.

So $36 - 33 = 3$ students scored more than 75%.

The interquartile range is

$64\% - 42\%$
$= 22\%$

Draw the lower quartile at cumulative frequency $= \dfrac{36}{4}$

The lower quartile was 42%.

There were 36 students in the class. (This is the FIRST FACT you should establish.)

Draw the upper quartile at cumulative frequency $= \dfrac{3 \times 36}{4}$

The upper quartile was 64%.

Draw the median at cumulative frequency $= \dfrac{36}{2}$

The median was 54%.

Now try this

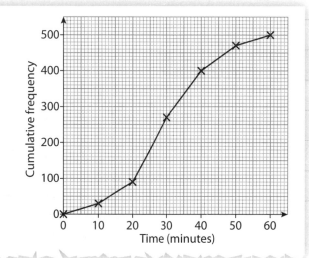

The cumulative frequency graph shows the journey times to college of 500 students.

(a) Write down the median time taken.
(1 mark)

(b) Work out the interquartile range of these times. *(2 marks)*

(c) What percentage of students took more than 44 minutes to travel to college?
(2 marks)

Box plots

Box plots show the median, upper and lower quartiles, and the largest and smallest values of a set of data. They are often used to compare distributions.

Half the weights were between 60 kg and 78 kg.
25% of the weights were greater than 78 kg.

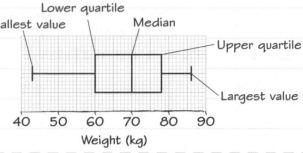

Smallest value   Lower quartile   Median   Upper quartile   Largest value

Weight (kg)

Drawing box plots

You might need to draw a box plot from data presented:

- as a list of values
- on a stem-and-leaf diagram
- in a cumulative frequency diagram.

To draw a box plot you ALWAYS need to find the LEAST and GREATEST values, the QUARTILES and the MEDIAN.

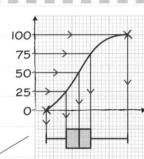

You can read the median and quartiles off a cumulative frequency diagram.

 Worked example

The box plot gives information about the heights, in metres, of the trees in a park.

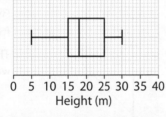

Height (m)

There are 162 trees in a park. Estimate the number of trees which are less than 10 m tall.
(1 mark)

$$12.5\% \text{ of } 175 = \frac{12.5}{100} \times 162 = 20.25$$

20 is a good estimate.

10 m is half-way between the smallest value and the lower quartile.

25% of the data values lie between the smallest value and the lower quartile.

So a good estimate is that 12.5% of the trees are less than 10 m tall.

Box plot checklist

Drawn on graph paper	✓
Ruler and sharp pencil	✓
Needs a scale	✓
Shows range and interquartile range	✓
Shows median and quartiles	✓

 Now try this

Here is a table showing information about some test scores for 120 boys.

(a) Draw a box plot for this data.
(3 marks)

(b) How many boys scored over 50?
(1 mark)

(c) Estimate how many boys scored less than 20.
(1 mark)

Lowest score	10
Lower quartile	29
Median	37
Interquartile range	21
Range	53

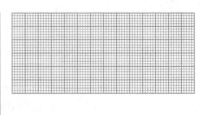

A*
A
B
C
D

Comparing data

You can use averages like the MEAN or MEDIAN and measures of spread like the RANGE and INTERQUARTILE RANGE to compare two sets of data. Follow these steps:

1 Calculate an average and a measure of spread for both data sets.

2 Write a sentence for each statistic COMPARING the values for each data set.

3 Only make a statement if you can back it up with STATISTICAL EVIDENCE.

Calculate then compare

You will often have to compare two sets of data presented in different ways. Make sure you calculate the SAME statistics for both data sets. You will get marks for calculating the statistics correctly AND for comparing the data sets.

Worked example

These box plots show information about the prices of used cars at two different garages.

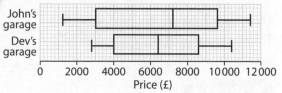

John's garage
Dev's garage

0 2000 4000 6000 8000 10000 12000
Price (£)

Compare the prices of used cars at the two garages.
(2 marks)

	Median (£)	IQR (£)
John's	7200	9600 − 3000 = 6600
Dev's	6400	8600 − 4000 = 4600

The cars at Dev's garage were cheaper on average (smaller median).

The prices of the cars at John's garage were more spread out (larger IQR).

Remember you need to mention one average **and** one measure of spread. You should use the **interquartile range** as your measure of spread if the data is presented on a box plot or cumulative frequency diagram.

Start by writing the median and IQR for each set of data. Then write a sentence comparing each statistic.

Golden rules

1 When you are comparing two sets of data you should use ONE AVERAGE like the median, and ONE MEASURE OF SPREAD like the interquartile range.

2 INTERPRET your results in the CONTEXT of the question.

Now try this

(a) The cumulative frequency graph shows some information about the times taken for 40 girls to complete a challenge. The shortest time taken was 8 minutes. Draw a box plot for this data. *(3 marks)*

(b) The box plot shows some information about the times taken by 40 boys to complete the same challenge.

Make **two** comments to compare the times taken by the girls and the times taken by the boys to complete the challenge.
(2 marks)

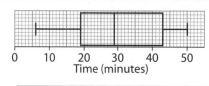

0 10 20 30 40 50
Time (minutes)

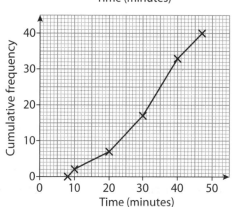

40
30
20
10
0

Cumulative frequency

0 10 20 30 40 50
Time (minutes)

Problem-solving practice 1

About half of the questions on your exam will need problem-solving skills.

These skills are sometimes called AO2 and AO3.

Practise using the questions on the next two pages.

For these questions you might need to:

- choose which mathematical technique or skill to use
- apply a technique in a new context
- plan your strategy to solve a longer problem
- show your working clearly and give reasons for your answers.

---

 *Callum is comparing two mobile phone packages.

> **Business First**
> £18 per month + 15% VAT
> 10% DISCOUNT FOR DIRECT DEBIT

> **Leisure One**
> One-off yearly payment of £260.
> 20% DISCOUNT FOR DIRECT DEBIT

Callum says that 'Business First' is cheaper if he pays by Direct Debit. Is Callum correct?
Show all your working.

(5 marks)

Percentage change 1 p.2   target **D**

You need to plan your answer and show all your working. Start by calculating the cost of each plan if Callum pays by Direct Debit. Then write a sentence comparing the costs and say whether Callum is correct.

TOP TIP

If a question has a * next to it then one mark is awarded for **quality of written communication**. This means you must show all your working and write your answer clearly with the correct units.

 Rania measured the reaction times of her class using a computer program. This frequency polygon shows her result.

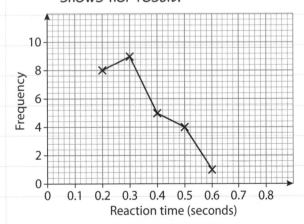

Reaction time (seconds)

Calculate an estimate of the mean reaction time.   *(3 marks)*

Frequency table averages p.13   target **C**

Frequency polygons are usually used to represent **grouped continuous data**. In a frequency polygon the values are plotted at the **midpoints** of the class intervals, so you have less work to do when estimating the mean.

TOP TIP

Make sure you can read and work with data represented in a table, a graph or as a list of numbers. In your exam you might be given data in any of these formats.

Problem-solving practice 2

3 At a sixth form college, the ratio of male to female students is 7 : 6. There are 144 female students at the college. How many students are there in total?

(2 marks)

Ratio p.4

Use this information in the question to work out what each part of the ratio represents. There are 7 + 6 = 13 parts in the ratio, so multiply your answer by 13 to work out the total number of students.

TOP TIP

Check your answer by dividing it in the ratio 7 : 6.

---

4 A council wants to take a sample of 100 primary school students. The sample will be selected from four schools, stratified by the size of each school. This table shows some information about the sample.

School	Number of students	Number in sample
Bounds green		
Chestnuts	394	26
Mulberry	679	
Campsbourne		12

Complete the table. *(4 marks)*

Sampling p.11

The row for 'Chestnuts' has both values given. Use this to work out the sampling proportion.

$$\frac{26}{394} = 0.0659\ldots$$

So the sample size is 6.6% (2 s.f.) of the population size for each school. You also need to use the fact that the total sample size is 100 to complete the table.

TOP TIP

There is sometimes more than one correct value for a question. If you use the correct method and show all your working you will get the marks.

---

5 Beccy has two bags of numbered balls.

Bag A   Bag B

She chooses a ball at random from bag A and places it in bag B. Then she chooses a ball at random from bag B and places it in bag A.

Calculate the probability that the sum of the balls in bag A will be 18 or greater. *(4 marks)*

Probability 1 p.18

Work out all the ways Beccy can end up with a total of 18 or more in bag A.
Use a table to keep track of your working:

Work out the probability of each outcome and add them together.

A → B	B → A	Total in A
1	4	18
1	5	19
2	5	18

TOP TIP

You can answer lots of tricky probability questions by writing down all the successful outcomes then working out the probability of each one.

25

Factors and primes

The FACTORS of a number are any numbers that divide into it exactly.

A PRIME NUMBER has exactly two factors. It can only be divided by 1 and itself.

Prime factors

If a number is a factor of another number AND it is a prime number then it is called a PRIME FACTOR. You use a factor tree to find prime factors.

Remember to circle the prime factors as you go along. The order doesn't matter.

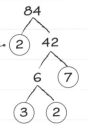

$84 = 2 \times 2 \times 3 \times 7$ → Remember to put in the multiplication signs.

$= 2^2 \times 3 \times 7$ → This is called a PRODUCT of PRIME FACTORS.

The highest common factor (HCF) of two numbers is the HIGHEST NUMBER that is a FACTOR of both numbers.

The lowest common multiple (LCM) of two numbers is the LOWEST NUMBER that is a MULTIPLE of both numbers.

Worked example

(a) Write 108 as the product of its prime factors. Give your answer in index form. *(2 marks)*

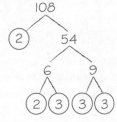

$108 = 2 \times 2 \times 3 \times 3 \times 3 = 2^2 \times 3^3$

(b) Work out the highest common factor (HCF) of 108 and 24. *(2 marks)*

$108 = ②\times②\times③\times 3 \times 3$
$24 = ②\times②\times 2 \times③$
HCF is $2 \times 2 \times 3 = 12$

(c) Work out the lowest common multiple (LCM) of 108 and 24. *(2 marks)*

$LCM = 12 \times 3 \times 3 \times 2 = 216$

Draw a factor tree. Continue until every branch ends with a prime number. This question asks you to write your answer in **index form**. This means you need to use **powers** to say how many times each prime number occurs in the product.

Check it!

$2^2 \times 3^3 = 4 \times 27 = 108$ ✓

To find the HCF circle all the prime numbers which are **common** to both products of prime factors. 2 appears twice in both products so you have to circle it twice. Multiply the circled numbers together to find the HCF.

To find the LCM multiply the HCF by the numbers in both products that were not circled in part **(b)**. You could also multiply 108 and 24 together and divide by the HCF:

$108 \times 24 = 2592$
$2592 \div 12 = 216$

Now try this

Use a factor tree.

(a) Write 180 as a product of prime factors. *(2 marks)*
(b) Work out the Highest Common Factor (HCF) of 180 and 630. *(2 marks)*
(c) Work out the Lowest Common Multiple (LCM) of 180 and 630. *(2 marks)*

A*
A
B
C
D

Fractions and decimals

You need to be able to work with decimals, fractions and mixed numbers WITHOUT a calculator for your Unit 2 exam.

Adding or subtracting fractions

Convert any mixed numbers to improper fractions

↓

Write all fractions as equivalent fractions with the same denominator

↓

Add or subtract the fractions

↓

Simplify your answer

Worked example

target **C**

(a) Work out

$2\frac{1}{3} + 3\frac{1}{6}$ *(2 marks)*

$\frac{7}{3} + \frac{19}{6} = \frac{14}{6} + \frac{19}{6}$

$= \frac{33}{6}$

$= \frac{11}{2}$

$= 5\frac{1}{2}$

(b) Work out

$4\frac{1}{4} - \frac{3}{5}$ *(2 marks)*

$\frac{17}{4} - \frac{3}{5} = \frac{85}{20} - \frac{12}{20}$

$= \frac{73}{20}$

$= 3\frac{13}{20}$

Converting between fractions and decimals

You can write a terminating decimal as a fraction with denominator 10, 100 or 1000.

$0.24 = \frac{24}{100} = \frac{6}{25}$ Simplify your fraction as much as possible.

To convert a fraction into a decimal you divide the numerator by the denominator.

$\frac{2}{5} = 2 \div 5 = 0.4$

It's useful to remember these common fraction-to-decimal conversions:

Fraction	$\frac{1}{100}$	$\frac{1}{20}$	$\frac{1}{10}$	$\frac{1}{2}$	$\frac{1}{5}$	$\frac{1}{4}$	$\frac{3}{4}$
Decimal	0.01	0.05	0.1	0.5	0.2	0.25	0.75

Worked example

target **D**

Write 0.27, $\frac{7}{20}$ and 32% in order with the smallest first.   *(2 marks)*

$\frac{1}{20} = 0.05$ so $\frac{7}{20} = 7 \times 0.05 = 0.35$

32% = 0.32

The order is 0.27, 32%, $\frac{7}{20}$

Remember that you can't use a calculator for this question. Learning the common fraction-to-decimal conversions will really help with questions like this.

Write all three numbers as decimals then compare them.

Now try this

target **D**

1 Write 0.6, $\frac{9}{20}$ and 65% in order, with the smallest first.   *(3 marks)*

2 Work out $\frac{3}{10} + \frac{5}{8}$ *(2 marks)*

target **C**

3 Work out $5\frac{1}{4} - 2\frac{4}{5}$ *(3 marks)*

Write both mixed numbers as improper fractions first.

27

A*
A
B
C
D

Decimals and estimation

You can APPROXIMATE the answer to a calculation by rounding each number to 1 significant figure and then doing the calculation.

This is useful for checking your answers.

$$4.32 \times 18.09 \approx 4 \times 20$$

This symbol means 'is approximately equal to'.

4.32 rounded to 1 significant figure is 4.

18.09 rounded to 1 significant figure is 20.

The calculation is approximately equal to 80.

Quick division trick!

If you multiply both numbers in a division by the same amount the answer stays the same.

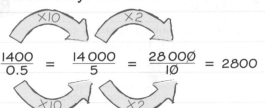

$$\frac{1400}{0.5} = \frac{14\,000}{5} = \frac{28\,00\emptyset}{1\emptyset} = 2800$$

Worked example

target C

Use approximations to estimate the value of

(a) $\dfrac{1.9 \times 740}{0.48}$　　　　(3 marks)

$$\frac{1.9 \times 740}{0.48} \approx \frac{2 \times 700}{0.5} = \frac{1400}{0.5} = 2800$$

(b) $\dfrac{\sqrt{97.2}}{0.24}$　　　　(2 marks)

$$\frac{\sqrt{97.2}}{0.24} \approx \frac{\sqrt{100}}{0.2} = \frac{10}{0.2} = 50$$

EXAM ALERT!

Do **not** round each number to 1 decimal place. If you are approximating an answer you should round each number to 1 significant figure:

$1.9 \to 2$　　$740 \to 700$　　$0.48 \to 0.5$

To divide by a fraction without a calculator you can use the quick division trick in the box above.

Students have struggled with exam questions similar to this – **be prepared!**

You might need to use the information given in a question to work out the answer to a calculation. You can use inverse operations, and check your answer by approximating.

Worked example

target D

Given that $3.6 \times 54 = 194.4$

(a) work out $\dfrac{19.44}{54}$　　　　(1 mark)

0.36

Check: $\dfrac{19.44}{54} \approx \dfrac{20}{50} = 0.4$ ✓

(b) work out 360×5.4　　　　(1 mark)

1944

Check: $360 \times 5.4 \approx 400 \times 5$
　　　　　　　　$= 2000$ ✓

Look at the calculation which is given.

(a) By inverse operations $194.4 \div 54 = 3.6$. The numerator has been divided by 10 and the denominator is unchanged. So you need to divide the answer by 10.

(b) 3.6 has been multiplied by 100 and 54 has been divided by 10:

$$\boxed{\times 100} \boxed{\div 10} \to \text{is the same as} \to \boxed{\times 10} \to$$

So you need to multiply the answer by 10.

Now try this

Remember to round all your values to 1 significant figure.

1　You are given that $24.6 \times 15.5 = 381.3$

Work out the answers to these calculations,

target D

(a) 2.46×1.55　　(b) $\dfrac{3813}{24.6}$

(c) $\dfrac{246 \times 155}{3813}$　　　　(3 marks)

target C

2　Use approximations to estimate the value of $\dfrac{201.8 \times 5.94}{0.297}$

(3 marks)

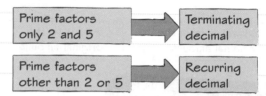

-A*-
-A-
-B-
-C-
-D-

Recurring decimals

A TERMINATING decimal can be written exactly as a decimal.

RECURRING decimals have one digit or group of digits repeated forever. You can use dots to show the recurring digit or group of digits.

$$\frac{2}{3} = 0.6666\ldots = 0.\dot{6}$$

The dot tells you that the 6 repeats forever.

$$\frac{346}{555} = 0.623\,4234\ldots = 0.6\dot{2}3\dot{4}$$

These dots tell you that the group of digits 234 repeats forever.

Recurring or terminating?

To check whether a fraction produces a recurring decimal or a terminating decimal, write it in its simplest form and find the prime factors of its DENOMINATOR.

Prime factors only 2 and 5	→	Terminating decimal

Prime factors other than 2 or 5	→	Recurring decimal

Worked example

(a) Show that $\frac{7}{50}$ can be written as a terminating decimal. *(1 mark)*

$$\frac{7}{50} = \frac{14}{100} = 0.14$$

(b) Show that $\frac{11}{24}$ can **not** be written as a terminating decimal. *(2 marks)*

$$\frac{11}{24} = \frac{11}{2^3 \times 3}$$

The denominator contains a factor other than 2 or 5 so the decimal is recurring.

You could also write the denominator as a product of prime factors:

$$\frac{7}{50} = \frac{7}{2 \times 5^2}$$

The only factors are 2 and 5 so $\frac{7}{50}$ produces a terminating decimal.

To convert a fraction to a decimal you divide the numerator by the denominator. You could also use long division to work out $2 \div 9$:

```
     0.2 2 2...
  9)2.0 0 0...
    1.8
    0.2 0 0
    0.1 8
    0.0 2 0
    0.0 1 8
```

Worked example

(a) Show that $\frac{2}{9}$ is equivalent to 0.222… *(1 mark)*

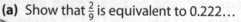
```
     0.2 2 2...
  9)2.²0²0²0...
```

(b) Hence, or otherwise, write 0.7222… as a fraction. *(3 marks)*

$$0.7222\ldots = 0.222\ldots + 0.5$$
$$= \frac{2}{9} + \frac{1}{2}$$
$$= \frac{4}{18} + \frac{9}{18} = \frac{13}{18}$$

Now try this

1 Which one of $\frac{5}{8}$, $\frac{7}{12}$, and $\frac{3}{10}$ is a recurring decimal? Show clearly how you made your decision. *(2 marks)*

2 Write $\frac{5}{11}$ as a recurring decimal. *(2 marks)*

Percentage change 2

For your Unit 2 exam, you need to be able to calculate a percentage increase or decrease without a calculator.

Work out the percentage

DECREASE → Subtract it from the original amount

INCREASE → Add it to the original amount

Calculating percentages

You can use multiples of 1% and 10% to calculate percentages without a calculator.

Work out 12.5% of £600

10% of £600 is £60	600 ÷ 10 = 60
1% of £600 is £6	600 ÷ 100 = 6
0.5% of £600 is £3	6 ÷ 2 = 3

So, 12.5% of £600 is
£60 + £6 + £6 + £3 = £75

Worked example

 target **D**

Which television is cheaper?
Show all of your working. *(5 marks)*

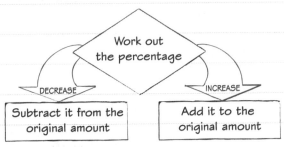

A £440 +20% VAT

B £550 SALE 2.5% OFF

Television A
10% = £44
20% = £88
£440 + £88 = £528

Television B
1% = £5.50
0.5% = £2.75
£550 − £5.50 − £5.50 − £2.75
= £536.25
Television A is cheaper.

You need to show **all your working**.

Television A

1. Divide by 10 to calculate 10% of £440.
2. Multiply by 2 to work out 20%.
3. Add this to the original price.

Television B

1. Divide by 100 to calculate 1% of £550.
2. Divide by 2 to work out 0.5%.
3. Subtract two lots of £5.50 and one lot of £2.75 from the original price.

Compare the prices

It's always a good idea to write your answer as a sentence.

Percentage increase and decrease questions come in lots of different forms.
Look out for these in your Unit 2 exam:

PERCENTAGE INCREASE

INTEREST
FIXED RATE
ISA 2.8%

PAY RISE

PERCENTAGE DECREASE

SALES

 10% off

DEPRECIATION

VALUE / TIME

Now try this

 target **D**

Simon sees the same model of digital camera for sale in two different shops.

Which shop is selling the camera at the cheaper price?

You **must** show your working. *(5 marks)*

CRUKS CAMERAS
30% OFF NORMAL PRICE OF £245

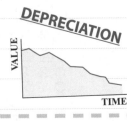

Spivs Cameras
35% OFF NORMAL PRICE OF £270

A*
A
B
C
D

Ratio problems

Ratio is used in lots of exam questions. Practise these questions WITHOUT a calculator – remember that you won't have one in your Unit 2 exam.

Worked example

target C

Tom, Jamie and Chaaya took part in a sponsored swim to raise money for charity.

The ratio of Tom's total to Jamie's total is 5 : 7.

Tom raised £12 less than Jamie.

Chaaya raised £9 more than Jamie.

How much money did they each raise?   *(3 marks)*

2 parts = £12 so 1 part = £6

Tom = £30

Jamie = £42

Chaaya = £51

Golden rule

You can answer lots of ratio questions by working out what ONE PART of the ratio represents.

Tom raised £12 less than Jamie so 2 parts of the ratio represents £12.

So Tom raised 5 × £6 = £30

Jamie raised 7 × £6 = £42

Chaaya raised £9 more than Jamie so she raised £51.

Worked example

target B

A paint manufacturer mixes white and green paint in two different ratios to make different shades.

APPLE WHITE

RATIO
WHITE : GREEN
5 : 2

SUMMER FIELDS

RATIO
WHITE : GREEN
1 : 4

Rosa has 2 litres of Summer Fields. How much extra white paint should she add to turn it into Apple White?   *(4 marks)*

2 litres ÷ 5 = 0.4 litres

There are 0.4 litres of white and 1.6 litres of green in 2 litres of Summer Fields.

1.6 litres ÷ 2 = 0.8 litres

0.8 litres × 5 = 4 litres

Rosa needs 4 litres of white paint in total, so she needs to add 3.6 litres of white paint.

EXAM ALERT!

This question has lots of steps. It's a good idea to **plan** your answer.

1. Divide 2 litres in the ratio 1 : 4 to work out how much white and green paint Rosa has.
2. The amount of green paint represents 2 parts in the **new** ratio. Divide by 2 to work out how much one part is worth.
3. Multiply by 5 to work out how much white paint is needed in total.
4. Subtract the amount of white paint Rosa already has to get your answer.

Check it!

If Rosa adds 3.6 litres she will have 5.6 litres in total. Divide 5.6 in the ratio 5 : 2.

5.6 ÷ 7 = 0.8

White = 0.8 × 5 = 4 litres ✓

Green = 0.8 × 2 = 1.6 litres ✓

Students have struggled with exam questions similar to this – **be prepared!**

Now try this

target C

Chris, Dave and Ed share £800 between them. Chris receives £128 more than Dave.

The ratio of Chris's share to Dave's share is 5 : 3. Work out the ratio of Ed's share to Dave's share.

Give your answer in its simplest form.

(5 marks)

Indices 1

Use these rules about indices to simplify calculations.

 Index laws

Indices include square roots, cube roots and powers.

You can use the index laws to simplify powers and roots.

$a^m \times a^n = a^{m+n}$

$4^3 \times 4^7 = 4^{3+7} = 4^{10}$

$\dfrac{a^m}{a^n} = a^{m-n}$

$12^8 \div 12^3 = 12^{8-3} = 12^5$

$(a^m)^n = a^{mn}$

$(7^3)^5 = 7^{3 \times 5} = 7^{15}$

 Cube root

The cube root of a positive number is positive.

$4 \times 4 \times 4 = 64$

$4^3 = 64$

$\sqrt[3]{64} = 4$

The cube root of a negative number is negative.

$-4 \times -4 \times -4 = -64$

$(-4)^3 = -64$

$\sqrt[3]{-64} = -4$

 Powers of 0 and 1

Anything raised to the power 0 is equal to 1.

$6^0 = 1 \quad 1^0 = 1 \quad 7223^0 = 1 \quad (-5)^0 = 1$

Anything raised to the power 1 is equal to itself.

$8^1 = 8 \quad 499^1 = 499 \quad (-3)^1 = -3$

Indices checklist

The base numbers have to be the same.

If there's no index, the number has the power 1

Be careful with negatives:
$(-3)^2 = 9$

Worked example

 target B

Here are two numbers.

A 7.5×10^{-1}　　**B** $\dfrac{2^2}{\sqrt[3]{125}}$

Which is the larger?
You **must** show your working.　(3 marks)

$7.5 \times 10^{-1} = 0.75$

$\dfrac{2^2}{\sqrt[3]{125}} = \dfrac{4}{5} = 0.8$

0.8 is larger than 0.75 so **B** is larger.

> You need to write both numbers as decimals. Show all the stages of your working, and remember you can't use a calculator for this question. For a reminder about numbers in **standard form** look at page 5.

Learn it!

You need to know the square numbers up to 15^2 and the cubes of 1, 2, 3, 4, 5 and 10. You also need to know the corresponding square roots and cube roots.

Now try this

target C

1 (a) Work out $2^3 \times \sqrt[3]{1000}$　(2 marks)

(b) Work out the possible values of x if $3x^2 = 48$　(2 marks)

> Divide both sides by 3 to get x^2 on its own. There are two possible answers.

target C

2 (a) Simplify $12x^0$　(1 mark)

(b) Simplify $(12x)^0$　(1 mark)

(c) Simplify $x^8 \times x^4$　(1 mark)

(d) Simplify $y^{20} \div y^5$　(1 mark)

Indices 2

Here are six key facts to help you answer indices questions in your Unit 2 exam.

A*
A
B
C
D

 Negative powers

$$a^{-n} = \frac{1}{a^n}$$

$$5^{-2} = \frac{1}{5^2} = \frac{1}{25}$$

Be careful!

A NEGATIVE power can still have a POSITIVE answer.

 Reciprocals

$$a^{-1} = \frac{1}{a}$$

This means that a^{-1} is the RECIPROCAL of a. You can find the reciprocal of a fraction by turning it upside down.

$$\left(\frac{5}{9}\right)^{-1} = \frac{9}{5}$$

 Powers of fractions

$$\left(\frac{a}{b}\right)^n = \frac{a^n}{b^n}$$

$$\left(\frac{3}{10}\right)^2 = \frac{3^2}{10^2} = \frac{9}{100}$$

4 **Combining rules**

You can apply the rules one at a time.

$$\left(\frac{a}{b}\right)^{-n} = \frac{b^n}{a^n}$$

$$\left(\frac{2}{3}\right)^{-3} = \left(\frac{3}{2}\right)^3 = \frac{3^3}{2^3} = \frac{27}{8}$$

5 **Fractional powers**

You can use fractional powers to represent roots.

$$a^{\frac{1}{2}} = \sqrt{a} \qquad 49^{\frac{1}{2}} = \pm 7$$

$$a^{\frac{1}{3}} = \sqrt[3]{a} \qquad 27^{\frac{1}{3}} = 3$$

$$a^{\frac{1}{4}} = \sqrt[4]{a} \qquad 16^{\frac{1}{4}} = \pm 2$$

Check it!

A whole number raised to a power less than 1 gets smaller.

6 **More complicated indices**

You can use the index laws to work out more complicated fractional powers.

$$a^{-\frac{m}{n}} = \left(\frac{1}{a^n}\right)^m$$

Do these calculations ONE STEP AT A TIME.

$$27^{-\frac{2}{3}} = (27^{\frac{1}{3}})^{-2}$$
$$= (\sqrt[3]{27})^{-2}$$
$$= 3^{-2} = \frac{1}{3^2} = \frac{1}{9}$$

 Worked example

target **A***

Find the value of n when $3^n = 9^{-\frac{3}{2}}$

(3 marks)

$$9^{-\frac{3}{2}} = (3^2)^{-\frac{3}{2}}$$
$$= 3^{2 \times -\frac{3}{2}} = 3^{-3}$$
So $3^n = 3^{-3}$ and $n = -3$

3^n is not the same as $3n$. You can't divide by 3 to get n on its own. You need to make the base on the right-hand side the same as the base on the left-hand side.

1. Write 9 as a power of 3. Remember to use brackets.
2. Use $(a^n)^m = a^{nm}$ to write the right-hand side as a single power of 3.
3. Compare both sides and write down the value of n.

 Now try this

target **A**

1. You are given that $x = 7^h$ and $y = 7^k$

Write each of the following as a single power of 7.

(a) $\frac{x}{y}$ *(1 mark)*

(b) x^2 *(1 mark)*

(c) xy^2 *(2 marks)*

 target **A***

2. Work out the value of $81^{-\frac{3}{4}}$ *(2 marks)*

3. Work out $4^{2.5} \div 64^{-0.5}$ *(3 marks)*

Write both numbers as powers of 2.

Surds

You can give exact answers to calculations by leaving some numbers as square roots.

? | Area = 10 cm²

This square has a side length of $\sqrt{10}$ cm.

You can't write $\sqrt{10}$ exactly as a decimal number. It is called a SURD.

Rules for simplifying square roots

These are the most important rules to remember when dealing with surds.

$$\sqrt{ab} = \sqrt{a} \times \sqrt{b} \qquad \sqrt{8} = \sqrt{4} \times \sqrt{2} = 2\sqrt{2}$$

$$\sqrt{\frac{a}{b}} = \frac{\sqrt{a}}{\sqrt{b}} \qquad \sqrt{\frac{3}{25}} = \frac{\sqrt{3}}{\sqrt{25}} = \frac{\sqrt{3}}{5}$$

You need to remember these rules for your exam. They are NOT given on the formula sheet.

Worked example

target A

Write $\sqrt{45}$ in the form $k\sqrt{5}$ where k is an integer. *(2 marks)*

$$\sqrt{45} = \sqrt{9 \times 5}$$
$$= \sqrt{9} \times \sqrt{5}$$
$$= 3\sqrt{5}$$

$$k = 3$$

Do **not** leave an answer as a square root if it can be written as an integer.

1. Look for a factor of 45 which is a square number: $45 = 9 \times 5$.
2. Use the rule $\sqrt{ab} = \sqrt{a} \times \sqrt{b}$ to split the square root into two square roots.
3. Write $\sqrt{9}$ as a whole number.

RATIONALISING THE DENOMINATOR of a fraction means making the denominator a whole number.

You can do this by multiplying the top AND bottom of the fraction by the surd part in the denominator.

$$\overset{\times\sqrt{2}}{\frac{5}{3\sqrt{2}}} = \frac{5\sqrt{2}}{6} \underset{\times\sqrt{2}}{}$$

The surd part of the denominator is $\sqrt{2}$

Remember that $\sqrt{2} \times \sqrt{2} = 2$
So $3\sqrt{2} \times \sqrt{2} = 3 \times 2 = 6$

Good form

Most surd questions ask you to write a number or answer in a certain FORM.

This means you need to find INTEGERS for all the letters in the expression.

$6\sqrt{3}$ is in the form $k\sqrt{3}$.
$$k = 6$$
The integers can be positive or negative.

$4 - 9\sqrt{2}$ is in the form $p + q\sqrt{2}$.
$$p = 4 \text{ and } q = -9$$

You can check your answer by writing down the integer value for each letter.

Now try this

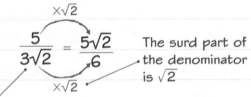

Start by writing each number as a multiple of $\sqrt{2}$.

target A

1 **(a)** Write $\sqrt{32} + \sqrt{98}$ in the form $p\sqrt{2}$ where p is an integer. *(2 marks)*

(b) Simplify $\frac{35}{\sqrt{7}}$ by rationalising the denominator. *(2 marks)*

target A*

2 Work out the value of x if
$$\frac{\sqrt{x} \times \sqrt{18}}{\sqrt{3}} = 8\sqrt{3}$$ *(4 marks)*

Algebraic expressions

-A*-
-A-
-B-
-C-
-D-

When tackling the most demanding questions, you need to make sure you can work with algebraic expressions confidently.

1 You can use the INDEX LAWS to simplify algebraic expressions.

$a^m \times a^n = a^{m+n}$

$x^4 \times x^3 = x^{4+3} = x^7$

$\dfrac{a^m}{a^n} = a^{m-n}$

$m^8 \div m^2 = m^{8-2} = m^6$

$(a^m)^n = a^{mn}$

$(n^2)^4 = n^{2 \times 4} = n^8$

2 You can square or cube a whole expression.

$(4x^3y)^2 = (4)^2 \times (x^3)^2 \times (y)^2$
$= 16x^6y^2$

You need to square everything inside the brackets.

$16 = (4)^2$

$(x^3)^2 = x^{3 \times 2} = x^6$

Remember that if a letter appears on its own then it has the power 1.

3 Algebraic expressions may also contain negative and fractional indices.

$a^{-m} = \dfrac{1}{a^m}$

$(c^2)^{-3} = c^{2 \times -3} = c^{-6} = \dfrac{1}{c^6}$

$a^{\frac{1}{n}} = \sqrt[n]{a}$

$(8p^3)^{\frac{1}{3}} = (8)^{\frac{1}{3}} \times (p^3)^{\frac{1}{3}}$
$= \sqrt[3]{8} \times p^{3 \times \frac{1}{3}}$
$= 2p$

One at a time

When you are MULTIPLYING expressions:

1. Multiply any number parts first.

2. Add the powers of each letter to work out the new power.

$$6p^2q \times 3p^3q^2 = 18p^5q^3$$

$6 \times 3 = 18$

$p^2 \times p^3 = p^{2+3} = p^5$

$q \times q^2 = q^{1+2} = q^3$

When you are DIVIDING expressions:

1. Divide any number parts first.

2. Subtract the powers of each letter to work out the new power.

$12 \div 3 = 4$

$$\dfrac{12a^5b^3}{3a^2b^2} = 4a^3b$$

$b^3 \div b^2 = b^{3-2} = b$

$a^5 \div a^2 = a^{5-2} = a^3$

 Worked example

target **C-B**

Simplify fully

(a) $(n^3)^3$ *(1 mark)*

$(n^3)^3 = n^9$

(b) $\dfrac{4n^5 \times 5n^2}{10n^3}$ *(3 marks)*

$\dfrac{4n^5 \times 5n^2}{10n^3} = \dfrac{20n^7}{10n^3} = 2n^4$

'Simplify' and 'simplify fully' mean you have to write each expression with a single number part and a single power of n.

(a) Use $(a^m)^n = a^{mn}$

(b) Start by simplifying the top part of the fraction. Do the number part first then the powers. Use $a^m \times a^n = a^{m+n}$

Next divide the expressions. Divide the number part then subtract the indices. You are using $\dfrac{a^m}{a^n} = a^{m-n}$

Now try this

target **C**

target **B**

1 Simplify $(h^2)^6$ *(1 mark)*

2 Simplify fully

   (a) $(2a^5b)^4$ *(2 marks)*

   (b) $5x^4y^2 \times 3x^3y^7$ *(2 marks)*

   (c) $18d^8g^{10} \div 6d^2g^5$ *(2 marks)*

target **A**

3 Simplify $(25p^6)^{\frac{1}{2}}$ fully.   *(2 marks)*

Remember you need to apply the power outside the brackets to everything inside the brackets.

A*
A
B
C
D

Expanding brackets

Expanding or multiplying out brackets is a key algebra skill.

You have to multiply the expression outside the bracket by everything inside the bracket.

$4n \times n^2 = 4n^3$

$$4n(n^2 + 2) = 4n^3 + 8n$$

$4n \times 2 = 8n$

'Expand and simplify' means 'multiply out and then collect like terms'.

Golden rule

When you expand, you need to be careful with negative signs in front of the bracket.

Negative signs belong to the term to their right.

$-2 \times x$　　$-2 \times -y$

$$x - 2(x - y) = x - 2x + 2y$$

$$= -x + 2y$$

Multiply out the brackets first and then collect like terms if possible.

You can use the GRID METHOD to expand two brackets.

$(x + 7)(x - 5) = x^2 - 5x + 7x - 35$

$$= x^2 + 2x - 35$$

Remember to collect like terms if possible.

	x	−5
x	x^2	−5x
7	7x	−35

The negative sign belongs to the 5.

You need to write it in your grid.

OR

You can use the acronym FOIL to expand two brackets.

$2a^2$　　$-b^2$

$$(2a + b)(a - b) = 2a^2 - 2ab + ab - b^2$$

$$= 2a^2 - ab - b^2$$

ab

$-2ab$

Some people remember this as a 'smiley face'.

First terms
Outer terms
Inner terms
Last terms

Worked example

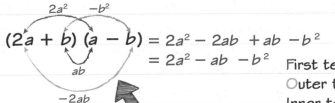

target B

Expand and simplify $(3p - 4)^2$　　(2 marks)

$(3p - 4)^2 = (3p - 4)(3p - 4)$

$$= 9p^2 - 12p - 12p + 16$$

$$= 9p^2 - 24p + 16$$

	3p	−4
3p	$9p^2$	−12p
−4	−12p	16

EXAM ALERT!

The question is 'expand and simplify' so you have to multiply out **and** collect like terms.

Use the grid method or FOIL to find all four terms of the expansion.

Be extra careful with your negative signs:

$-4 \times -4 = 16$　　$p \times -4 = -4p$

Students have struggled with exam questions similar to this – **be prepared!**

Now try this

target D

1　Expand $5(a - 6)$　　(2 marks)

2　Expand and simplify

　(a) $3(2b - 1) - 4(b + 2)$　　(2 marks)

target C

　(b) $8y(3y^3 + 4)$　　(2 marks)

　(c) $2x(x^2 + 6) + 3x(x - 5)$　　(3 marks)

　(d) $(m + 9)(m - 3)$　　(2 marks)

target A

3　Expand and simplify

　(a) $(3g - 4e)(2g - 5e)$　　(2 marks)

　(b) $(4x + 7)^2$　　(2 marks)

$(4x + 7)^2 = (4x + 7)(4x + 7)$

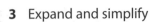

A*
A
B
C
D

Factorising

Factorising is the opposite of expanding brackets:

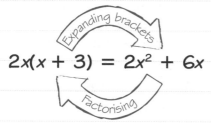

$$2x(x + 3) = 2x^2 + 6x$$

You need to look for the LARGEST FACTOR you can take out of every term in the expression.

$$10a^2 + 5ab = 5(2a^2 + ab)$$

This expression has only been PARTLY FACTORISED.

$$10a^2 + 5ab = 5a(2a + b)$$

This expression has been COMPLETELY FACTORISED.

Factorising $x^2 + bx + c$

You need to write the expression with TWO BRACKETS.

You need to find two numbers which add up to 7... $5 + 2 = 7$

$$x^2 + 7x + 10 = (x + 5)(x + 2)$$

... and multiply to make 10 $5 \times 2 = 10$

When factorising $x^2 + bx + c$, use this table to help you find the two numbers:

b	c	Factors
Positive	Positive	Both numbers positive
Positive	Negative	Bigger number positive and smaller number negative
Negative	Negative	Bigger number negative and smaller number positive
Negative	Positive	Both numbers negative

Factorising $ax^2 + bx + c$

$$2x^2 - 7x - 15 = (2x\quad)(x\quad)$$

One of the brackets must contain a $2x$ term. Try pairs of numbers which have a product of -15. Check each pair by multiplying out the brackets.

$(2x + 5)(x - 3) = 2x^2 - x - 15$ ✗
$(2x - 3)(x + 5) = 2x^2 + 7x - 15$ ✗
$(2x + 3)(x - 5) = 2x^2 - 7x - 15$ ✓

Difference of two squares

You can factorise expressions that are written as
$$(\text{something})^2 - (\text{something else})^2$$
Use this rule:
$$a^2 - b^2 = (a + b)(a - b)$$
$$x^2 - 36 = x^2 - 6^2$$
$$= (x + 6)(x - 6)$$
36 is a square number.
$36 = 6^2$ so $a = x$ and $b = 6$

Worked example

Factorise fully
(a) $p^2 - 3p$ **(b)** $15x^2 + 5xy$
    (1 mark)   *(2 marks)*

$p(p - 3)$ $5x(3x + y)$

You need to look for the **largest** factor you can take out of every term.
Partly factorised: $x(15x + 5y)$ ✗
Partly factorised: $5(3x^2 + xy)$ ✗
Fully factorised: $5x(3x + y)$ ✓

Now try this

1 Factorise
   (a) $4a - 6$ *(1 mark)*
   (b) $y^2 + 5y$ *(1 mark)*

2 Factorise fully
   (a) $12g + 3g^2$ *(2 marks)*
   (b) $p^2 - 15p + 14$ *(2 marks)*
   (c) $6x^2 - 8xy$ *(2 marks)*

3 Factorise
   (a) $4ma - 24m^2a$ *(2 marks)*
   (b) $p^2 - 64$ *(1 mark)*

4 Factorise fully
   (a) $3y^2 + 7y - 20$ *(2 marks)*
   (b) $x^2 - 81y^2$ *(2 marks)*
   (c) $50g^2 - 2e^2$ *(3 marks)*

Algebraic fractions

Simplifying an algebraic fraction is just like simplifying a normal fraction.

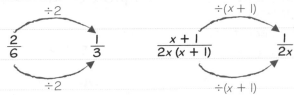

$$\frac{2}{6} \xrightarrow{\div 2} \frac{1}{3}$$

$$\frac{x+1}{2x(x+1)} \xrightarrow{\div(x+1)} \frac{1}{2x}$$

You can divide the top and bottom by a number, a term, or a whole expression.

Golden rule

If the top or the bottom of the fraction has MORE THAN one term, you will need to factorise before simplifying.

$$\frac{p^2 + 3p}{4p} = \frac{p(p+3)}{4p} = \frac{p+3}{4}$$

Two terms on top so factorise the top, then divide the top and bottom by p.

Operations on algebraic fractions

To ADD or SUBTRACT algebraic fractions with different denominators:

1. Find a common denominator.
2. Add or subtract the numerators.
3. Simplify if possible.

The smallest common denominator isn't always the product of the two denominators.

$$\frac{1}{x+4} + \frac{2}{x-4} = \frac{x-4}{(x+4)(x-4)} + \frac{2(x+4)}{(x+4)(x-4)}$$

$$= \frac{x-4+2x+8}{(x+4)(x-4)} = \frac{3x+4}{(x+4)(x-4)}$$

You can use a common denominator of $4x$ to simplify this expression:

$$\frac{x+1}{2x} + \frac{3-2x}{4x}$$

2 To MULTIPLY fractions:

1. Multiply the numerators AND multiply the denominators.
2. Simplify if possible.

$$\frac{x}{2} \times \frac{4}{x-1} = \frac{\overset{2}{\cancel{4}}x}{\underset{1}{\cancel{2}}(x-1)} = \frac{2x}{x-1}$$

Don't expand brackets if you don't have to. It's much easier to simplify your fraction with the brackets in place.

3 To DIVIDE fractions:

1. Change the second fraction to its reciprocal.
2. Change ÷ to ×.
3. Multiply the fractions and simplify.

$$\frac{x^2}{3} \div \frac{x}{6} = \frac{x^2}{3} \times \frac{6}{x} = \frac{\overset{2}{\cancel{6}}x^{\cancel{2}}}{\underset{1}{\cancel{3}}x} = 2x$$

To find the reciprocal of a fraction you turn it upside down.

Worked example

 target A*

Simplify $\dfrac{3x^2 - 8x - 3}{x^2 - 9}$ (3 marks)

$$\frac{3x^2 - 8x - 3}{x^2 - 9} = \frac{(3x+1)(x-3)}{(x+3)(x-3)}$$

$$= \frac{3x+1}{x+3}$$

EXAM ALERT!

You need to factorise the top and the bottom of the fraction before you can simplify. Remember that $a^2 - b^2 = (a+b)(a-b)$.

Students have struggled with exam questions similar to this – **be prepared!**

Now try this

target A

1 Simplify fully

(a) $\dfrac{a+1}{3a} + \dfrac{7}{6a}$ (3 marks)

(b) $\dfrac{3y}{(y-3)(y+6)} - \dfrac{2}{y+6}$ (4 marks)

 target A*

2 Simplify fully

(a) $\dfrac{x^2 + 4x - 12}{x^2 - 25} \div \dfrac{x+6}{x^2 - 5x}$ (5 marks)

(b) $\dfrac{3m^2 - 108}{9m^3 + 54m^2}$ (3 marks)

Straight-line graphs

Here are three things you need to know about straight-line graphs.

1 If an equation is in the form $y = mx + c$, its graph will be a straight line.

$$y = -\tfrac{1}{2}x + 5$$

This number tells you the gradient of the graph.

The y-intercept of the graph is at $(0, 5)$.

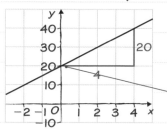

The gradient of the graph is $-\tfrac{1}{2}$.

This means that for every unit you go across, you go half a unit down.

2 Use a table of values to draw a graph.

$y = 2x + 1$

x	-1	0	1	2
y	-1	1	3	5

$y = 2 \times 2 + 1 = 5$

Choose simple values of x and substitute them into the equation to find the values of y.

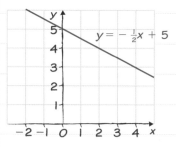

Plot the points on your graph and join them with a straight line.

---

3 When tackling the most demanding questions, you need to be able to find the equation of a line.

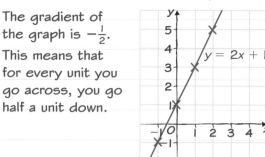

Draw a triangle to find the gradient.

$$\text{Gradient} = \frac{20}{4} = 5$$

The y-intercept is at $(0, 20)$.

Put your values for m and c into the equation of a straight line, $y = mx + c$.

The equation is $y = 5x + 20$.

Straight-line checklist

Plot at least three points when drawing straight-line graphs.

Lines of the form $y = mx$ pass through $(0, 0)$.

Lines of the form $y = mx + c$ pass through the y-axis at $(0, c)$. ✓

m is positive, line slopes up. ✓

m is negative, line slopes down.

Worked example **target D**

On the grid, draw the graph of $x + y = 3$.
Use values of x from -2 to 4.　　*(3 marks)*

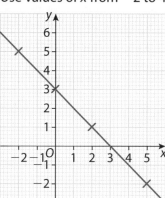

x	-2	0	2	5
y	5	3	1	-2

Everything in red is part of the answer.

Now try this **target D-C**

(a) Draw a graph of $y = 5 - 2x$
　　(3 marks)

(b) Write down the gradient and the y-intercept of this straight line.
　　(2 marks)

Gradients and midpoints

PARALLEL lines have the same gradient.
These three lines all have a gradient of 1.

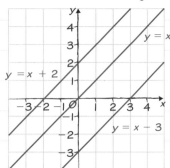

$y = x$
$y = x + 2$
$y = x - 3$

Parallel lines NEVER MEET.

Midpoints

A LINE SEGMENT is a short section of a straight line. You can find the MIDPOINT of a line segment if you know the coordinates of the ends.

$$\text{Coordinates of midpoint} = \left(\begin{array}{c} \text{average of} \\ x\text{-coordinates} \end{array}, \begin{array}{c} \text{average of} \\ y\text{-coordinates} \end{array} \right)$$

$$\frac{-3 + 5}{2} \qquad \frac{8 + (-3)}{2}$$

$(-3, 8) \times$ ⟶ $(1, 2\frac{1}{2})$

Midpoint ⟶ $\times$ (5, −3)

Worked example

target B

On the grid, M is the midpoint of the line segment PQ. The gradient of PQ is $\frac{1}{2}$.

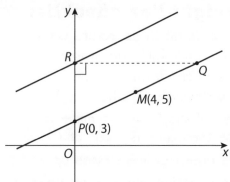

R

Q

$M(4, 5)$

$P(0, 3)$

Work out the equation of the line through R that is parallel to PQ.   **(3 marks)**

$5 - 3 = 2 \qquad 5 + 2 = 7$

The y-coordinate of Q and R is 7

The line has equation $y = \frac{1}{2}x + 7$

You need to work out the y-coordinate of Q. You know that M is the midpoint so the vertical distance from P to M is the same as the distance from M to Q.

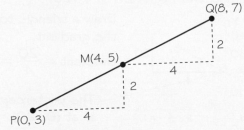

$Q(8, 7)$

2

$M(4, 5)$

4

2

$P(0, 3)$ 4

R is level with Q so it has the same y-coordinate. Because the line through R is parallel to PQ it has the same gradient, $\frac{1}{2}$. You know the gradient and y-intercept of the line so you can write down its equation using $y = mx + c$.

Now try this

target A*

The graph shows two lines A and B.

The equation of line A is $5x + 2y + 10 = 0$.

Line A crosses the axes at P and Q.

Q is the mid-point of PR.

Line B crosses the x-axis at the point S (4, 0).

Line B crosses the y-axis at T.

R is the mid-point of ST.

Work out the equation of line B.   **(6 marks)**

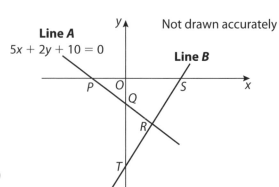

Line A
$5x + 2y + 10 = 0$

Not drawn accurately

Line B

P O Q S x

R

T

A*-A-A-B-C-D

Real-life graphs

You can draw graphs to explain real-life situations. This graph shows the cost of buying some printed T-shirts from three different companies.

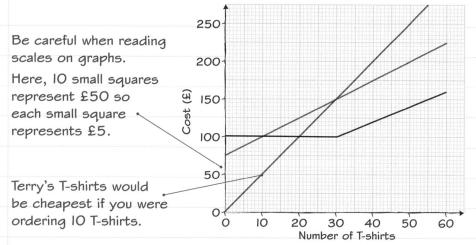

Be careful when reading scales on graphs.

Here, 10 small squares represent £50 so each small square represents £5.

Terry's T-shirts would be cheapest if you were ordering 10 T-shirts.

TERRY'S T-SHIRTS
No minimum order
£5 per shirt

SHIRT-O-GRAPH
Set-up cost £75
£2.50 per shirt

PAM'S PRINTING
Special offer 30 shirts for £100
Additional shirts just £2

A DISTANCE–TIME graph shows how distance changes with time. This distance–time graph shows Jodi's run. The shape of the graph gives you information about the journey.

A horizontal line means no movement. Jodi rested here for 15 minutes.

The horizontal scale might be marked in minutes or hours. Remember that there are 60 minutes in 1 hour.

At 13:15 Jodi was 1.4 miles from home.

The gradient of the graph gives Jodi's speed.

$$\text{Gradient} = \frac{\text{distance up}}{\text{distance across}} = 1.9 \div \frac{1}{2} = 3.8$$

Jodi was travelling at 3.8 mph on this section of the run.

1.9 miles

½ hour

This is when Jodi turned around and started heading back home.

Straight lines mean that Jodi was travelling at a constant speed.

Jodi sped up when she was nearly home. The graph is steeper here.

Now try this

D-C

Beth leaves home in her car at 09:30 and returns at 11:45.

The graph shows her journey.

(a) How far does she travel altogether? *(1 mark)*

(b) For how long does the car stop altogether? *(2 marks)*

(c) Work out the speed of the car on the fastest part of the journey. *(3 marks)*

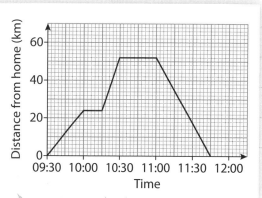

A*
A
B
C
D

Formulae

A FORMULA is a mathematical rule.

You can write formulae using algebra.

This label shows a formula for working out the cooking time of a chicken.

FREE-RANGE CHICKEN		
WEIGHT (KG)	PRICE PER KG	COOKING INSTRUCTIONS
1.8	£3.95	Cook at 170°C for 25 minutes per kg plus half an hour

You can write this formula using algebra as

$T = 25w + 30$, where T is the cooking time in minutes and w is the weight in kg.

In the description of each variable, you must give the units.
If T was the cooking time in hours then this formula would give you a very crispy chicken!

Worked example

This formula is used in physics to calculate distance.

$D = ut - 5t^2$

$u = 14$ and $t = -3$

Work out the value of D.   (2 marks)

$D = (14)(-3) - 5(-3)^2$

$\quad = (14)(-3) - 5(9)$

$\quad = -42 - 45$

$\quad = -87$

Substitute the values for u and t into the formula.

If you use brackets then you're less likely to make a mistake. This is really important when there are negative numbers involved.

Remember **BIDMAS** for the correct order of operations. You need to do:

Indices → Multiplication → Subtraction

Don't try to do more than one operation on each line of working.

Worked example

Larry's Limousines uses this graph to calculate the cost of a journey of m miles.

Circle the correct formula for the cost, C.   (1 mark)

$\boxed{C = 4m + 10}$ $C = 14m$

$C = 10m + 2$ $C = 4m$

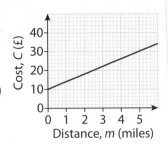

Look at the graph. When $m = 0$, $C = 10$. The only formula which makes this true is the first one.

Check it!
When $m = 5$,
$C = 4 \times 5 + 10 = 30$ ✓

Now try this

The graph shows the cost of hiring a car from Bloggs Car Hire.

(a) Circle the correct formula connecting the cost, C, and the number of days, d.

$C = 30d + 100$ $C = 30d + 10$

$C = 10d + 30$ $C = 10d + 100$ (1 mark)

(b) The cost of hiring a car from Trips Car Hire is given by the formula $C = 8d + 40$.
On the axes given, plot the graph of $C = 8d + 40$.   (2 marks)

(c) For what number of days will the cost be the same for Bloggs and Trips Car Hire?   (1 mark)

A*
A
B
C
D

Linear equations 1

To solve a linear equation you need to get the letter on its own on one side.
It is really important to write your working NEATLY when you are solving equations.

$$5x + 3 = 18 \quad (-3)$$
$$5x = 15 \quad (\div 3)$$
$$x = 3$$

Write down the operation you are carrying out. Remember to do the same thing to both sides of the equation.

Every line of working should have an equals sign in it.

Start a new line for each step. Do one operation at a time.

Line up the equals signs.

Letter on both sides?

To solve an equation you have to get the letter on its own on one side of the equation.

Start by collecting like terms so that all the letters are together.

$$2 - 2x = 26 + 4x \quad (+ 2x)$$
$$2 = 26 + 6x \quad (- 26)$$
$$-24 = 6x \quad (\div 6)$$
$$-4 = x$$

You can write your answer as

$-4 = x$ or as $x = -4$

Equations with brackets

Always start by multiplying out the brackets then collecting like terms.

For a reminder about multiplying out brackets have a look at page 36.

$$19 = 8 - 2(5 - 3y)$$
$$19 = 8 - 10 + 6y$$
$$19 = -2 + 6y \quad (+ 2)$$
$$21 = 6y \quad (\div 6)$$
$$\frac{21}{6} = y$$
$$y = \frac{7}{2} \text{ or } 3\frac{1}{2} \text{ or } 3.5$$

Your answer can be written as a fraction or decimal.

Worked example

Solve $7r + 2 = 5(r - 4)$ (3 marks)

$$7r + 2 = 5r - 20 \quad (- 5r)$$
$$2r + 2 = -20 \quad (- 2)$$
$$2r = -22 \quad (\div 2)$$
$$r = -11$$

Multiply out the brackets then collect all the terms in r on one side.

Check it!

Substitute $r = -11$ into each side of the equation.

Left-hand side: $7(-11) + 2 = -75$

Right-hand side: $5(-11 - 4) = -75$ ✓

Now try this

1 Solve
   (a) $5w - 17 = 2w + 4$ (3 marks)
   (b) $2(x + 11) = 20$ (3 marks)

Expand the brackets first.

2 Solve
   (a) $6y - 9 = 2(y - 8)$ (3 marks)
   (b) $4m - 2(m - 3) = 7m - 14$ (3 marks)

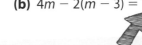

Expand the brackets then collect all the m terms on one side of the equation.

43

A* A B C D

Linear equations 2

Equations with fractions

When you have an equation with fractions, you need to get rid of any fractions before solving. You can do this by multiplying every term by the lowest common multiple (LCM) of the denominators.

$$\frac{x}{3} + \frac{x-1}{5} = 11 \qquad (\times 15)$$

The LCM of 3 and 5 is 15.

$$\frac{^5\cancel{15}x}{\cancel{3}_1} + \frac{^3\cancel{15}(x-1)}{\cancel{5}_1} = 165$$

Cancel the fractions. There is more about simplifying algebraic fractions on page 38.

$$5x + 3x - 3 = 165$$
$$8x - 3 = 165 \qquad (+3)$$
$$8x = 168 \qquad (\div 8)$$
$$x = 21$$

Multiplying by an expression

You might have to multiply by an expression to get rid of the fractions.

$$\frac{20}{n-3} = -5 \qquad (\times(n-3))$$
$$20 = -5(n-3)$$

Worked example

Solve $\dfrac{29-x}{4} = x + 5$ *(3 marks)*

$$\frac{4(29-x)}{4} = 4(x+5)$$
$$29 - x = 4(x+5)$$
$$29 - x = 4x + 20 \qquad (+x)$$
$$29 = 5x + 20 \qquad (-20)$$
$$9 = 5x \qquad (\div 5)$$
$$\frac{9}{5} = x$$

EXAM ALERT!

You need to get rid of any fractions **before** you start solving the equation. Multiply both sides of the equation by 4. Use brackets to show that you are multiplying everything by 4.

$$4(x+5) \checkmark \qquad 4x + 5 ✗$$

Multiply out the brackets and solve normally. Your answer can be a whole number or a fraction.

Check it!
If you have time, check each line of your working carefully.

> Students have struggled with exam questions similar to this – **be prepared!**

Writing your own equations

You can find unknown values by writing and solving equations.

$4(x-1)$ cm $(3x+3)$ cm

$$4(x-1) = 3x + 3$$

$\frac{5}{n}$ m

Perimeter = 20 m   2 m

$$\frac{5}{n} + \frac{5}{n} + 2 + 2 = 20$$

Now try this

1 Solve

(a) $\dfrac{25 - 3w}{4} = 10$ *(3 marks)*

(b) $5x - 10 = \dfrac{18 - x}{3}$ *(3 marks)*

2 Solve

(a) $\dfrac{2y}{3} + \dfrac{y-4}{2} = 5$ *(3 marks)*

(b) $\dfrac{3m-1}{4} - \dfrac{2m+4}{3} = 1.5$ *(3 marks)*

Number machines

A number machine is sometimes called a FUNCTION MACHINE. It takes an INPUT and applies OPERATIONS to it to produce an OUTPUT. You might need to write equations to solve problems involving number machines in your Unit 2 exam.

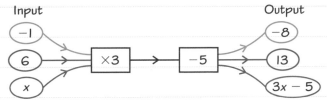

Using algebra

If you don't know the input for a number machine you can write it as x. You can then use algebra to write the output as an expression.

Golden rule

If the output and the input of a number machine are BOTH BLANK you need to write an EQUATION to find the missing values.

Worked example

target C

Here is a number machine:

Input → ×5 → −2 → Output

The output is equal to the input.
Work out the input.   *(3 marks)*

INPUT	OUTPUT
x	$5x - 2$

$$5x - 2 = x$$
$$5x = x + 2$$
$$4x = 2$$
$$x = \tfrac{1}{2}$$

EXAM ALERT!

Read the question very carefully. It will tell you the **relationship** between the input and the output.

Students have struggled with exam questions similar to this – **be prepared!**

Follow these steps when the output and input are both blank.

- Write the input as x.
- Work out an expression for the output.
- Use the information in the question to write an equation.
- Solve your equation to work out the value of x.

Check it!
$\tfrac{1}{2} \times 5 - 2 = 2\tfrac{1}{2} - 2 = \tfrac{1}{2}$ ✓

Now try this

target C

Here is a number machine.

Input → +4 → ×3 → Output

The output is 5 more than the input.
Work out the input.   *(3 marks)*

Inequalities

An inequality tells you when one value is bigger or smaller than another value.
You can represent INEQUALITIES on a number line.

$x > -1$

−3 −2 −1 0 1 2 3 4

Use an OPEN circle for > and <

The open circle shows that −1 is NOT included.

$x \leqslant 3$

−3 −2 −1 0 1 2 3 4

Use a CLOSED circle for ⩾ and ⩽

The closed circle shows that 3 IS included.

Solving inequalities

You can solve an inequality in exactly the same way as you solve an equation.

$x - 3 \leqslant 12$ (+ 3)

$x \leqslant 15$

The solution has the letter on its own on one side of the inequality and a number on the other side.

Golden rule

If you MULTIPLY or DIVIDE both sides of an inequality by a NEGATIVE number you have to REVERSE the INEQUALITY sign.

$6 - 5x > 10$ (− 6)

$-5x > 4$ (÷ −5)

$x < -\frac{4}{5}$

You have divided by a negative number so you have to reverse the inequality sign.

Worked example

Solve $8x - 7 > 3x + 3$ (2 marks)

$8x > 3x + 10$ (+ 7)
$5x > 10$ (− 3x)
$x > 2$ (÷ 5)

EXAM ALERT!

This is an **inequality** and not an equation. You must not use an '=' sign in your answer. Remember that the solution has the letter on its own on one side and a number on the other side.

Students have struggled with exam questions similar to this – **be prepared!**

Integer solutions

You might need to write down all the integer solutions of an inequality.
INTEGERS are positive or negative whole numbers, including 0.

$-3 \leqslant x < 2$

−4 −3 −2 −1 0 1 2 3 4 5

This shows that x is between −3 and 2. It can equal −3 but cannot equal 2.

The integer solutions to this inequality are −3, −2, −1, 0 and 1.

Now try this

1 *n* is an integer. List the values of *n* such that $-9 < 3n \leqslant 6$.   (2 marks)

2 *n* is an integer. List the values of *n* such that $0 \leqslant n + 4 < 5$.   (2 marks)

3 Solve the inequality $3n + 8 > 2$.   (2 marks)

4 Solve the inequality $2(n - 5) \geqslant n + 12$ (2 marks)

Start by multiplying out the brackets. Then subtract n from both sides to get the letter on its own.

46

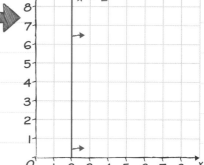

Inequalities on graphs

You can show the points that satisfy inequalities involving x and y on a graph.

For example, follow these steps to shade the region R that satisfies the inequalities:

$$x \geqslant 2 \qquad y > x \qquad x + y \geqslant 6$$

Always work on one inequality at a time.

1

$x \geqslant 2$

Draw the graph of $x = 2$ with a solid line.

Use a small arrow to show which side of the line you want.

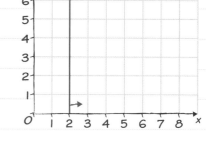

2

$y > x$

Draw the graph of $y = x$ with a dotted line.

Show which side of the line you want.

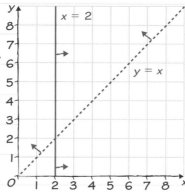

3

$x + y \geqslant 6$

Draw the graph of $x + y = 6$ with a solid line. Use a table of values.

x	0	3	6
y	6	3	0

Show which side of the line you want.

$x + y$ increases as you move away from the origin.

Shade in the region and label it **R**.

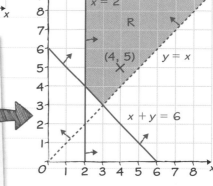

Graphical inequalities checklist

< and > are shown by DOTTED lines. ✓

≤ and ≥ are shown by SOLID lines. ✓

Points on a solid line ARE included in the region. ✓

Points on a dotted line ARE NOT included in the region. ✓

EXAM ALERT!

Always check your answer

Pick a point inside your shaded region.

Check that the x- and y-values for that point satisfy **all** the inequalities.

At (4, 5) $x = 4$ and $y = 5$.

$x \geqslant 2$ ✓ $y > x$ ✓ $x + y \geqslant 6$ ✓

> Students have struggled with exam questions similar to this – **be prepared!**

Now try this

On a grid, shade the region satisfied by these three inequalities:

$$y > 1 \qquad y \leqslant 2x + 1 \qquad 2x + 3y \leqslant 12$$

(4 marks)

Simultaneous equations 1

Simultaneous equations have two unknowns. You need to find the values for the two unknowns which make BOTH equations true.

Algebraic solution

1. Number each equation.

2. If necessary, multiply the equations so that the coefficients of one unknown are the same.

3. Add or subtract the equations to ELIMINATE that unknown.

4. Once one unknown is found use substitution to find the other.

5. Check the answer by substituting both values into the original equations.

$$3x + y = 20 \quad (1)$$
$$x + 4y = 14 \quad (2)$$
$$12x + 4y = 80 \quad (1) \times 4$$
$$- (x + 4y = 14) \quad - (2)$$
$$\overline{\qquad\qquad\qquad}$$
$$11x = 66$$
$$x = 6$$

Substitute $x = 6$ into (1):
$$3(6) + y = 20$$
$$18 + y = 20$$
$$y = 2$$

Solution is $x = 6$, $y = 2$.

Check: $x + 4y = 6 + 4(2) = 14$ ✓

Worked example

Solve the simultaneous equations

$$6x + 2y = -3 \quad (1)$$
$$4x - 3y = 11 \quad (2) \quad \textbf{(4 marks)}$$

$$18x + 6y = -9 \quad (1) \times 3$$
$$+ \ 8x - 6y = 22 \quad (2) \times 2$$
$$\overline{\qquad\qquad\qquad}$$
$$26x = 13$$
$$x = \tfrac{1}{2}$$

Substitute $x = \tfrac{1}{2}$ into (1):
$$6\left(\tfrac{1}{2}\right) + 2y = -3$$
$$3 + 2y = -3$$
$$2y = -6$$
$$y = -3$$

Easier eliminations

You can save time by choosing the right unknown to eliminate. Look for one of these:

 If an unknown appears ON ITS OWN in one equation you only need to multiply one equation to eliminate that unknown.

 If an unknown has DIFFERENT SIGNS in the two equations you can eliminate by ADDING.

Multiply both equations by a whole number to make the coefficients of y the same.

Check it!
Always use the equation you **didn't** substitute into to check your answer:
$$4x - 3y = 4\left(\tfrac{1}{2}\right) - 3(-3) = 2 + 9 = 11 \ ✓$$

Now try this

1 Solve the simultaneous equations

$$3x - 2y = 12$$
$$x + 4y = 11$$

You **must** show your working.
Do **not** use trial and improvement. *(3 marks)*

2 Solve the simultaneous equations

$$2x - 3y = 11$$
$$3x + 4y = 8$$

You **must** show your working.
Do **not** use trial and improvement. *(4 marks)*

3 Ali and Jen are buying pens and rulers.
Ali buys three pens and two rulers for 74p.
Jen buys five pens and one ruler for 79p.
Work out the cost of a pen and the cost of a ruler.
You **must** show your working.
Do **not** use trial and improvement. *(4 marks)*

Form two equations. Use x to represent the cost of one pen and y to represent the cost of one ruler.

A* A B C D

Quadratic equations

Quadratic equations can be written in the form $ax^2 + bx + c = 0$ where a, b and c are numbers.

To SOLVE a quadratic equation in your Unit 2 exam you need to:

1. REARRANGE it into the form $ax^2 + bx + c = 0$.

2. FACTORISE the left-hand side.

3. Set each factor EQUAL TO ZERO and solve to find two values of x.

For a reminder about factorising quadratic expressions have a look at page 37.

Two to watch

 When $c = 0$:
$$x^2 - 10x = 0$$
$$x(x - 10) = 0$$
Solutions are $x = 0$ and $x = 10$.

 When $b = 0$ (difference of two squares):
$$9x^2 - 4 = 0$$
$$(3x + 2)(3x - 2) = 0$$
Solutions are $x = \frac{2}{3}$ and $x = -\frac{2}{3}$.

Worked example

 target B

Solve $x^2 + 8x - 9 = 0$ *(3 marks)*

$$(x + 9)(x - 1) = 0$$

$x + 9 = 0$ $x - 1 = 0$

$x = -9$ $x = 1$

Follow the three steps given above.

1. The equation is already in the right form.

2. To factorise look for two numbers which add up to 8 and multiply to make -9. The numbers are 9 and -1.

3. Set each factor equal to 0 and solve.

Check it!
$(1)^2 + 8(1) - 9 = 1 + 8 - 9 = 0 ✓$
$(-9)^2 + 8(-9) - 9 = 81 - 72 - 9 = 0 ✓$

Worked example

 target A

Solve $2(x + 1)^2 = 3x + 5$ *(4 marks)*

$$2(x^2 + 2x + 1) = 3x + 5$$
$$2x^2 + 4x + 2 = 3x + 5$$
$$2x^2 + x - 3 = 0$$
$$(2x + 3)(x - 1) = 0$$

$2x + 3 = 0$ $x - 1 = 0$

$x = -\frac{3}{2}$ $x = 1$

EXAM ALERT!

When you are solving a quadratic equation you must rearrange it into the form $ax^2 + bx + c = 0$ **before** you factorise.

Be really careful if the coefficient of x^2 is bigger than 1. The two factors will look like this:

$$2x^2 + x - 3 = 0$$
$$(2x \pm \ldots)(x \pm \ldots) = 0$$

The number part of the expression is -3, so the numbers in the factors must be either -1 and 3 or 1 and -3.

Students have struggled with exam questions similar to this – **be prepared!**

Now try this

 target B

1. Solve $m^2 - 8m + 12 = 0$ *(3 marks)*

2. Solve $w^2 - 36 = 5w$ *(3 marks)*

target A

3. Solve $5y^2 + 37y - 24 = 0$ *(3 marks)*

4. Solve $8x^2 - 4 = (x - 1)^2$ *(4 marks)*

A*
A
B
C
D

Quadratics and fractions

You need to remove fractions before you can solve an equation.

For a reminder about solving linear equations with fractions have a look at page 44.

To remove fractions from an equation multiply everything by the lowest common multiple of the denominators.

$$\frac{x}{2x - 3} + \frac{4}{x + 1} = 1$$

→ $(2x - 3)$ and $(x + 1)$ don't have any common factors. Multiply everything by $(2x - 3)(x + 1)$.

$$\frac{x(2x-3)(x+1)}{2x-3} + \frac{4(2x-3)(x+1)}{x+1} = (2x-3)(x+1)$$

→ Don't expand brackets until you have simplified the fractions.

$$x(x + 1) + 4(2x - 3) = (2x - 3)(x + 1)$$

$$x^2 + x + 8x - 12 = 2x^2 - 3x + 2x - 3$$

$$0 = x^2 - 10x + 9$$

$$= (x - 9)(x - 1)$$

Multiply out brackets and collect like terms.

The solutions are $x = 9$ and $x = 1$.

Worked example

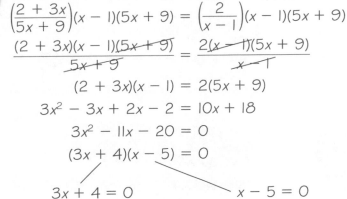

target A*

Solve $\dfrac{2 + 3x}{5x + 9} = \dfrac{2}{x - 1}$ (6 marks)

$$\left(\frac{2 + 3x}{5x + 9}\right)(x - 1)(5x + 9) = \left(\frac{2}{x - 1}\right)(x - 1)(5x + 9)$$

$$\frac{(2 + 3x)(x - 1)(5x + 9)}{5x + 9} = \frac{2(x - 1)(5x + 9)}{x - 1}$$

$$(2 + 3x)(x - 1) = 2(5x + 9)$$

$$3x^2 - 3x + 2x - 2 = 10x + 18$$

$$3x^2 - 11x - 20 = 0$$

$$(3x + 4)(x - 5) = 0$$

$$3x + 4 = 0 \qquad\qquad x - 5 = 0$$

$$x = -\frac{4}{3} \qquad\qquad x = 5$$

Multiply everything by $(x - 1)(5x + 9)$ to remove the fractions

If you are confident working with algebraic fractions you can jump straight to this step.

Set the two factors equal to 0 to find the two solutions.

Quadratic equations checklist

Remove any fractions by multiplying everything by the lowest common multiple of the denominators.   ✓

Multiply out any brackets and collect like terms.   ✓

Rewrite in the form $ax^2 + bx + c = 0$.   ✓

Factorise the left-hand side to solve the quadratic equation.   ✓

Now try this

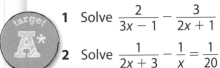

target A*

1 Solve $\dfrac{2}{3x - 1} - \dfrac{3}{2x + 1} = \dfrac{2}{5}$

→ Multiply everything by $(3x - 1)(2x + 1)$.   (6 marks)

2 Solve $\dfrac{1}{2x + 3} - \dfrac{1}{x} = \dfrac{1}{20}$ (6 marks)

A*
A
B
C
D

Simultaneous equations 2

If a pair of simultaneous equations involves an x^2 or a y^2 term, you need to solve them using SUBSTITUTION. Remember to NUMBER the equations to keep track of your working.

Rearrange the linear equation to make y the subject.

$$y = x^2 - 2x - 7 \qquad (1)$$
$$x - y = -3 \qquad (2)$$
$$y = x + 3 \qquad (3)$$
$$x + 3 = x^2 - 2x - 7 \qquad \longrightarrow \text{Substitute (3) into (1).}$$
$$0 = x^2 - 3x - 10$$
$$0 = (x - 5)(x + 2)$$

Each solution for x has a corresponding value of y. Substitute into (3) to find the two solutions.

$$x = 5 \text{ or } x = -2$$

The solutions are $x = 5$, $y = 8$ and $x = -2$, $y = 1$.

Worked example

Solve the simultaneous equations
$$x - 2y = 1 \qquad (1)$$
$$x^2 + y^2 = 13 \qquad (2) \qquad \text{(6 marks)}$$

$$x = 1 + 2y \qquad (3)$$
Substitute (3) into (2):
$$(1 + 2y)^2 + y^2 = 13$$
$$1 + 4y + 4y^2 + y^2 = 13$$
$$5y^2 + 4y - 12 = 0$$
$$(5y - 6)(y + 2) = 0$$
$$y = \frac{6}{5} \qquad \text{or} \quad y = -2$$
$$x = 1 + 2\left(\frac{6}{5}\right) \qquad x = 1 + 2(-2)$$
$$= \frac{17}{5} \qquad \qquad = -3$$
Solutions: $x = \frac{17}{5}$, $y = \frac{6}{5}$ and
$$x = -3, y = -2$$

EXAM ALERT!

In your Unit 2 exam, simultaneous equations involving y^2 or x^2 will always have **two pairs** of solutions. Each solution is an x-value **and** a y-value. You need to find **four** values in total and pair them up correctly.

Students have struggled with exam questions similar to this – **be prepared!**

You can substitute for x or y. It is easier to substitute for x because there will be no fractions.

Use brackets to make sure that the whole expression is squared.

Rearrange the quadratic equation for y into the form $ay^2 + by + c = 0$.

Factorise the left-hand side to find two solutions for y.

Thinking graphically

The solutions to a simultaneous equation correspond to the points where the graphs of each equation INTERSECT. Because an equation involving x^2 or y^2 represents a CURVE, there can be two points of intersection. Each point has an x-value and a y-value. You can write the solutions using coordinates.

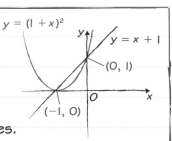

$y = (1 + x)^2$

$y = x + 1$

$(0, 1)$

$(-1, 0)$

Now try this

1 Solve the simultaneous equations
$$x - 2y = 3$$
$$x^2 + 2y^2 = 27$$
Do **not** use trial and improvement.
You **must** show your working. *(6 marks)*

2 Solve the simultaneous equations
$$2x = y + 1$$
$$y^2 = 7 + 9x - x^2$$
Do **not** use trial and improvement.
You **must** show your working. *(6 marks)*

Rearranging formulae

Most formulae have one letter on its own on one side of the formula. This letter is called the SUBJECT of the formula.

$$e = mc^2 \qquad e \text{ is the subject of the formula.}$$

CHANGING THE SUBJECT of a formula is like solving an equation. You have to do the same thing to both sides of the formula until you have the new letter on its own on one side.

$$e = mc^2 \qquad (\div m)$$
$$\frac{e}{m} = c^2 \qquad (\sqrt{\ })$$
$$\sqrt{\frac{e}{m}} = c$$

The inverse operation to x^2 is $\sqrt{\ }$. You need to square root EVERYTHING on both sides of the formula.

c is now the subject of the formula.

Harder formulae

If the letter you need APPEARS TWICE in the formula you need to FACTORISE.

| GROUP all the terms with that letter on one side of the formula and all the other terms on the other side. | → | FACTORISE so the letter only appears once. | → | DIVIDE by everything in the bracket to get the letter on its own. |

For a reminder about factorising have a look at page 37.

Worked example

$$N = \frac{3h + 20}{100}$$

Rearrange the formula to make h the subject. *(3 marks)*

$$N = \frac{3h + 20}{100} \qquad (\times 100)$$
$$100N = 3h + 20 \qquad (-20)$$
$$100N - 20 = 3h \qquad (\div 3)$$
$$\frac{100N - 20}{3} = h$$
$$h = \frac{100N - 20}{3}$$

It's a good idea to write your final answer as $h = \ldots$

Worked example

Make Q the subject of the formula $P = \dfrac{Q}{Q - 100}$

(4 marks)

$$P = \frac{Q}{Q - 100} \qquad [\times (Q - 100)]$$
$$P(Q - 100) = Q \quad \text{(multiply out brackets)}$$
$$PQ - 100P = Q \quad (+ 100P)$$
$$PQ = Q + 100P \quad (- Q)$$
$$PQ - Q = 100P \quad \text{(factorise)}$$
$$Q(P - 1) = 100P \qquad [\div (P - 1)]$$
$$Q = \frac{100P}{P - 1}$$

Your final answer should look like $Q = \ldots$
You need to factorise to get Q on its own.

Now try this

1 Rearrange this formula to make t the subject.

$$4p = 3t - 1 \qquad \text{(2 marks)}$$

2 Make w the subject of $m = \sqrt{5w + 7}$

(3 marks)

 Square both sides first.

3 Make y the subject of $x = \dfrac{5 - 9y}{y + 2}$

(4 marks)

Sequences 1

An ARITHMETIC SEQUENCE is a sequence of numbers where the difference between consecutive terms is CONSTANT. In your exam, you might need to work out the nth term of a sequence. Look at this example which shows you how to do it in four steps.

 1 Here is a sequence.

$$1 \;\boxed{+4}\; 5 \;\boxed{+4}\; 9 \;\boxed{+4}\; 13 \;\boxed{+4}\; 17$$

Work out a formula for the nth term of the sequence. *(2 marks)*

Write in the difference between each term.

2 Here is a sequence.

Zero term
$$-3 \quad 1 \;\boxed{+4}\; 5 \;\boxed{+4}\; 9 \;\boxed{+4}\; 13 \;\boxed{+4}\; 17$$

Work out a formula for the nth term of the sequence. *(2 marks)*

Work backwards to find the **zero term** of the sequence. You need to subtract 4 from the first term.

3 Here is a sequence.

Zero term
$$-3 \quad 1 \;\boxed{+4}\; 5 \;\boxed{+4}\; 9 \;\boxed{+4}\; 13 \;\boxed{+4}\; 17$$

Work out a formula for the nth term of the sequence. *(2 marks)*

nth term = difference $\times n$ + zero term

Write down the formula for the nth term.
Remember this formula for the exam.

4 Here is a sequence.

Zero term
$$-3 \quad 1 \;\boxed{+4}\; 5 \;\boxed{+4}\; 9 \;\boxed{+4}\; 13 \;\boxed{+4}\; 17$$

Work out a formula for the nth term of the sequence. *(2 marks)*

nth term = difference $\times n$ + zero term
nth term = $4n - 3$

Is 99 in this sequence?

You can use the nth term to check whether a number is a term in the sequence.

The value of n in your nth term has to be a POSITIVE whole number.

Try some different values of n:

$n = 25 \rightarrow 4n - 3 = 97$
$n = 26 \rightarrow 4n - 3 = 101$

You can't use a value of n between 25 and 26 so 99 is NOT a term in the sequence.

Check it!

Check your answer by substituting values of n into your nth term.
1st term: when $n = 1$,
$4n - 3 = 4 \times 1 - 3 = 1$ ✓
2nd term: when $n = 2$,
$4n - 3 = 4 \times 2 - 3 = 5$ ✓

You can also generate any term of the sequence.

For the 20th term, $n = 20$:
$4n - 3 = 4 \times 20 - 3 = 77$

So the 20th term is 77.

Now try this

Here is a sequence: 2 5 8 11 14 ......

(a) Work out an expression for the nth term of the sequence. *(2 marks)*
(b) Work out the 29th term in the sequence. *(2 marks)*
(c) How many terms of this sequence are **less than** 200? *(2 marks)*
(d) Is 156 a term in this sequence? *(2 marks)*

A*
A
B
C
D

Sequences 2

Generating sequences

You can work out the terms of a sequence by substituting the term number into the nth term. Here are two examples:

nth term	$n^2 + 10$	$2^n - 1$
1st term	$1^2 + 10 = 11$	$2^1 - 1 = 1$
2nd term	$2^2 + 10 = 14$	$2^2 - 1 = 3$
3rd term	$3^2 + 10 = 19$	$2^3 - 1 = 7$
⋮	⋮	⋮
8th term	$8^2 + 10 = 74$	$2^8 - 1 = 255$

Sequences and equations

You can use the nth term or the term-to-term rule of a sequence to write an EQUATION.

This sequence has term-to-term rule 'multiply by 2 then add a':

...   11   25   53   ...

So $2 \times 11 + a = 25$ and $a = 3$. You could use this information to find the next term in the sequence.

You know two consecutive terms so you can solve an equation to find the value of k.

Then use the term-to-term rule to find the 1st term. You could also use inverse operations.

$\boxed{-4}$ $\boxed{\times 3}$

8   12

$\boxed{+4}$ $\boxed{\div 3}$

The nth term of a sequence is $50 - n^2$.

(a) Work out the first three terms.   *(2 marks)*

   1st term $= 50 - 1^2 = 49$

   2nd term $= 50 - 2^2 = 46$

   3rd term $= 50 - 3^2 = 41$

(b) Work out the first term of the sequence that is negative.   *(2 marks)*

   7th term $= 50 - 7^2 = 1$

   8th term $= 50 - 8^2 = -14$

The 8th term is the first negative term.

Worked example   *target* **C**

The nth term of a sequence is $5(2n - 3)$

A term in the sequence is 75

Work out the value of n for this term. *(3 marks)*

   $5(2n - 3) = 75$

     $10n - 15 = 75$ $(+ 15)$

       $10n = 90$ $(\div 9)$

         $n = 9$

Worked example   *target* **A**

The rule for finding the next term in a sequence

is $\boxed{\text{Subtract } k \text{ then multiply by 3}}$

The second term is 12 and the third term is 24.

...   12   24   ...

Work out the 1st term of the sequence.

                   (4 marks)

   $24 = 3(12 - k)$

   $24 = 36 - 3k$

   $3k = 12$

    $k = 4$

If the first term is T then:

   $3(T - 4) = 12$

   $3T - 12 = 12$

      $3T = 24$

       $T = 8$

Now try this

target **D**

1 The nth term of a sequence is $4n + 27$.

   (a) Work out the first three terms.

                   (2 marks)

   (b) Work out which term of the sequence is the first one greater than 100.

                   (2 marks)

target **A**

2 The rule for finding the next term in a sequence is

  $\boxed{\text{Add } k \text{ and then multiply by 2}}$

The first four terms are a b 14   46   ...

Work out the values of a and b.   *(4 marks)*

A*
A
B
C
D

Algebraic proof

You can use algebra to PROVE facts about numbers.

Using algebra helps you to prove that something is true for EVERY number.

In a proof question the working IS the answer.

If the question says 'SHOW THAT...' or 'PROVE THAT...', you need to write down every stage of your working. If you don't, you won't get all the marks.

Golden rule

If you need to prove something about numbers then you always use algebra.

Algebraic proof toolkit

Use n to represent any whole number.

Number fact	Written using algebra
Even number	$2n$
Odd number	$2n + 1$ or $2n - 1$
Multiple of 3	$3n$
Consecutive numbers	$n, n + 1, n + 2, \ldots$
Consecutive even numbers	$2n, 2n + 2, 2n + 4, \ldots$
Consecutive odd numbers	$2n + 1, 2n + 3, 2n + 5, \ldots$
Consecutive square numbers	$n^2, (n + 1)^2, (n + 2)^2, \ldots$

Worked example

Prove that the sum of three consecutive integers is always divisible by 3. *(3 marks)*

$n, n + 1$ and $n + 2$ represent any three consecutive integers.

$n + (n + 1) + (n + 2) = 3n + 3$
$$= 3(n + 1)$$

$n + 1$ is an integer, so $3(n + 1)$ is divisible by 3.

To **prove** the statement you need to show that it is true for **any** three consecutive integers. You can do this using algebra.

1. Write the first integer as n and the next two integers as $n + 1$ and $n + 2$.
2. Write an expression for the sum of your three integers. Brackets can help to make your working clearer. Simplify your expression.
3. Factorise the expression.
4. Explain why the final expression is divisible by 3.

True or false?

To explain why something is TRUE for ANY NUMBER you need to use algebraic proof.

If you need to explain why something is FALSE you just need to write down one COUNTER-EXAMPLE. Here is an example:

STATEMENT: The sum of three consecutive numbers will always be even.

COUNTER-EXAMPLE: $2 + 3 + 4 = 9$

This *counter-example* shows that a statement is FALSE.

Now try this

1 Jill says that for any two prime numbers, a and b, $(2a + b)^2$ is always odd.

Give an example to show that she is wrong. *(2 marks)*

Write the two integers as n and $(n + 4)$.

2 Two integers have a difference of 4. The difference between the squares of the two integers is eight times the mean of the integers.

For example, $14 - 10 = 4$,

$14^2 - 10^2 = 196 - 100 = 96$,

The mean of 10 and 14 is $\frac{10 + 14}{2} = 12$,

and $8 \times 12 = 96$

Prove this result algebraically. *(4 marks)*

Identities

An IDENTITY is something which is ALWAYS true. The right-hand side of an identity is exactly equal to the left-hand side for any values of the variables. You use the symbol $\equiv$ to represent an identity.

Worked example

target A

Show that $(n - 1)^2 + (n + 1)^2 \equiv 2(n^2 + 1)$

(2 marks)

$(n - 1)^2 + (n + 1)^2$

$\equiv (n - 1)(n - 1) + (n + 1)(n + 1)$

$\equiv n^2 - 2n + 1 + n^2 + 2n + 1$

$\equiv 2n^2 + 2$

$\equiv 2(n^2 + 1)$

If you need to **show** that an identity is true in an exam you should show every line of your working.

Start with the expression on the left-hand side. Use multiplying out, simplifying and factorising to work towards the expression on the right-hand side.

Equating coefficients

You can find unknown values in an identity by comparing the left-hand side and the right-hand side.

- Coefficients of x^2 must be equal.
- Coefficients of x must be equal.
- Any number parts must be equal.

Golden rule

An identity is not like an equation. Do not solve it using the balance method.

Manipulate each side separately. ✓

Apply the same operation to both sides. ✗

Follow these steps to find a and b:

1. Expand the brackets on the left-hand side.
2. Group together the two x terms.
3. Write the coefficient of x on the left-hand side in terms of a: $ax + 20x \equiv (a + 20)x$
4. The coefficients of x on each side must be equal. Use this fact to write an equation and solve it to find a.
5. The number parts on each side must be equal. Use this fact and your value of a to calculate b.

Check it!

$(5x + 4)(x + 4) = 5x^2 + 4x + 20x + 16$

$= 5x^2 + 24x + 16$ ✓

Worked example

target A*

Here is an identity.

$(5x + a)(x + 4) \equiv 5x^2 + 6ax + b$

a and b are numbers.

Work out the values of a and b. (5 marks)

$5x^2 + ax + 20x + 4a \equiv 5x^2 + 6ax + b$

$5x^2 + (a + 20)x + 4a \equiv 5x^2 + 6ax + b$

Equate coefficients for x:

$a + 20 = 6a$

$20 = 5a$

$a = 4$

Equating number parts:

$4a = b$

$b = 16$

Now try this

target A*

Here is an identity:

$(4x + h)(x + k) \equiv 4x^2 + kx - 75$

h and k are integers.

Work out all possible pairs of values of h and k.
You must show your working. (5 marks)

You will need to write down two simultaneous equations and solve them to find h and k.

Completing the square

-A*-
-A-
-B-
-C-
-D-

If a quadratic expression is written in the form $(x + p)^2 + q$ it is in COMPLETED SQUARE form. You can solve quadratic equations which don't have integer answers by completing the square.

Useful identities

If you learn these two identities you can save time when you are completing the square:

1 $x^2 + 2bx + c \equiv (x + b)^2 - b^2 + c$

2 $x^2 - 2bx + c \equiv (x - b)^2 - b^2 + c$

Positive and negative roots

Remember that any positive number has TWO square roots: one POSITIVE and one NEGATIVE. If you 'square root' both sides of an equation you need to use ± (plus-or-minus) to show that there are two square roots.

$$x^2 = 4 \qquad (x + 4)^2 = 3$$
$$x = \pm 2 \qquad (x + 4) = \pm\sqrt{3}$$
$$x = -4 \pm\sqrt{3}$$

Worked example

(a) Find values of p and q such that
$$x^2 + 6x - 20 \equiv (x + p)^2 + q \qquad \text{(2 marks)}$$

$$x^2 + 6x - 20 = (x + 3)^2 - 3^2 - 20$$
$$= (x + 3)^2 - 9 - 20$$
$$= (x + 3)^2 - 29$$
$$p = 3 \text{ and } q = -29$$

Compare the expression with the identities for completing the square.
$$x^2 + 6x - 20 \equiv (x + p)^2 + q$$
$$x^2 + 2bx + c \equiv (x + b)^2 - b^2 + c$$
$2b = 6$ so $b = 3$

$c = -20$

Subsititute these values into the identity and simplify to find p and q.

(b) Hence, or otherwise, solve the equation
$$x^2 + 6x - 20 = 0$$
Give your answer in surd form.   (2 marks)

$$(x + 3)^2 - 29 = 0 \qquad (+29)$$
$$(x + 3)^2 = 29 \qquad (\sqrt{\ })$$
$$x + 3 = \pm\sqrt{29} \qquad (-3)$$
$$x = -3 \pm\sqrt{29}$$

Use your answer to part **(a)** to write the expression in completed square form. The unknown only appears once so you can solve it using **inverse operations**.

Remember to use the ± symbol when you take square roots of both sides. The two solutions are $x = -3 + \sqrt{29}$ and $x = -3 - \sqrt{29}$.

Now try this

1 Solve the equation
$$x^2 + 10x + 12 = 0$$
Give your answer in surd form.   (3 marks)

2 Solve the equation
$$y^2 - 6y - 15 = 0$$
Give your answer in the form $a \pm b\sqrt{c}$ where a, b and c are integers.   (4 marks)

Remember you can't use a calculator for this question. Start by writing $x^2 + 10x + 12$ in completed square form. Then you can solve the equation using inverse operations.

Problem-solving practice 1

About half of the questions on your exam will need problem-solving skills.

These skills are sometimes called AO2 and AO3.

Practise using the questions on the next two pages.

For these questions you might need to:

- choose which mathematical technique or skill to use
- apply a technique in a new context
- plan your strategy to solve a longer problem
- show your working clearly and give reasons for your answers.

---

1 Adam is selling honey at a farmers' market.

He begins the day with 90 jars of honey. He sells $\frac{4}{9}$ of them at £4.50 each. At lunchtime he reduces his prices by one-third. He sells $\frac{7}{10}$ of his remaining jars at the reduced price.

How much money has Adam made? **(5 marks)**

Fractions and decimals p.27

There are lots of stages here so you need to write down all your working to keep track. You could draw a table to show how many jars Adam sells at each price.

TOP TIP

Practise common fraction and percentage increases and decreases without a calculator. £4.50 ÷ 3 = £1.50 so Adam's new price after lunch is £3.00 per jar.

---

2 Jamie, Amir and Helen are comparing their ages.

Jamie is six years older than Amir.

Helen is three times older than Amir.

The sum of all three of their ages is 41 years.

Let x be Amir's age.

Set up and solve an equation to find Amir's age. **(3 marks)**

Linear equations 1 p.43

Write expressions for Jamie's age and Helen's age in terms of x. Add together all **three** expressions (including x for Amir's age) and set this new expression equal to 41. You can solve this equation to find x.

TOP TIP

If the question tells you to **use an equation**, do not use trial and improvement.

---

3 Here is an addition pyramid. Each expression is the sum of the two numbers underneath it.

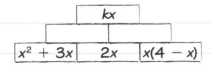

Work out the value of k. **(4 marks)**

Algebraic expressions p.35

Simplify each expression and fill in the second row of the table.
Here's one more entry:

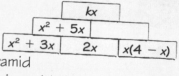

Work out an expression for the top box in the pyramid and compare your expression with kx.

TOP TIP

Don't panic if a question looks like it has lots of algebra. If you can do it with numbers, you can do it with letters!

Problem-solving practice 2

4 The diagram shows a line ABCD.
A is the point (30, 18).
B is the point (18, 12).
The line cuts the y-axis at C and the x-axis at D.

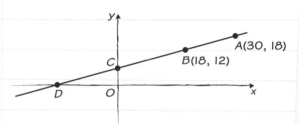

Work out the coordinates of C and D.

(4 marks)

Straight-line graphs p.39

You need to find the equation of the line. Draw a right-angled triangle with AB as its longest side to find the gradient. Then substitute the x and y values at one of the points into $y = mx + c$ and solve the equation to find c. Remember that the x- and y-values for any point on a line satisfy the equation of the line.

TOP TIP

When you are solving straight-line graph problems you might need to substitute values into $y = mx + c$ and solve an equation to find the gradient or the y-intercept.

---

5 Here is a number machine:

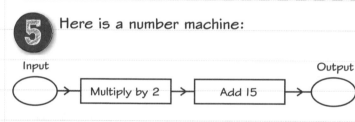

When the input is a the output is b.
When the input is b the output is c.
a is a positive integer, and $c = a^2$.
Work out the value of a.

(5 marks)

Number machines p.45
Quadratic equations p.49

Write an expression for b in terms of a. Use this expression as the next input for the number machine.

Write your expression for c equal to a^2. This gives you a quadratic equation with one positive solution and one negative solution. You are told that a is positive so you can ignore the negative solution.

TOP TIP

When substituting an expression always use brackets to make sure you don't make any mistakes:

$$c = 2b + 15 = 2(2a + 15) + 15$$

---

6 *The difference between the squares of two consecutive integers is equal to the sum of the two integers.

For example $9^2 - 8^2 = 81 - 64$
$$= 17$$

and $9 + 8 = 17$

Prove this result algebraically.

(3 marks)

Algebraic proof p.55

You can write the first integer as n and the second one as $n + 1$. Then the difference between their squares is $(n + 1)^2 - n^2$. You need to show that this is equal to $n + (n + 1)$.

TOP TIP

If a question has a * next to it there are marks available for **quality of written communication**. This means you need to write your answer clearly using the correct notation, and show all of your working.

A*
A
B
C
D

Proportion

In your Unit 3 exam you might need to answer word problems involving proportion.

Two quantities are in DIRECT PROPORTION when both quantities increase at the same rate.

Number of theatre tickets bought   Total cost

3   £135
×3   ×3
9   £405

Two quantities are in INVERSE PROPORTION when one quantity increases at the same rate as the other quantity decreases.

Average speed   Time taken

40 km/h   2 hours
×2   ÷2
80 km/h   1 hour

Worked example

target **D**

Suresh buys 4 picture frames for a total cost of £11.40.

Work out the cost of 7 of these picture frames.   *(2 marks)*

$$\text{Cost of 1 frame} = \frac{£11.40}{4} = £2.85$$

$$\text{Cost of 7 frames} = £2.85 \times 7 = £19.95$$

EXAM ALERT!

Calculate the cost of 1 picture frame first. Then multiply the cost of 1 frame by 7 to work out the cost of 7 frames. When you are working with money you should:
- do all your calculations in either £ or p
- write £ or p in your answer, but not both
- write answers in £ to 2 decimal places.

Students have struggled with exam questions similar to this – **be prepared!**

Divide or multiply?

6 people can build a wall in 4 days.
How long would it take 8 people to build the same wall?

Inverse proportion problems often involve time. The more people working on a task, the quicker it will be finished.

You can solve this problem by working out how long it would take 1 person to build the wall. Use common sense to decide whether to divide or multiply.

6 × 4 = 24 so 1 person could build the wall in 24 days.

You multiply because it would take 1 person more time to build the wall.

24 ÷ 8 = 3 so 8 people could build the wall in 3 days.

You divide because it would take 8 people less time to build the wall.

Now try this

target **D**

1 3 chairs cost £46.14 altogether.
How much would 5 of these chairs cost?   *(2 marks)*

Work out the cost of one chair.

target **D**

2 It takes 8 men a total of 6 days to dig a hole.
How long would it take three men to dig a hole of the same size?   *(2 marks)*

Work out how long it would take 1 man to dig the hole. It will take longer, so multiply.

Trial and improvement

A*
A
B
C
D

Some equations can't be solved exactly. You need to use trial and improvement to find an approximate solution. You will be told when to use trial and improvement in your exam. Look at this worked example which shows you how to do it in two steps.

Worked example

1 Use trial and improvement to find the solution to the equation

$$x^3 - 5x = 60$$

Give your answer correct to 1 decimal place.

x	x³ − 5x	Comment
4	44	Too low
5	100	Too high
4.5	68.625	Too high

(4 marks)

You will usually be given one trial in your exam, and a table to record your results.

x = 4 is too low, so try x = 5.

You can use the x^3 on your calculator to work out $x^3 - 5x$. Compare your answer with 60 and write down whether it is too high or too low.

x = 5 is too high. This means that the answer is between 4 and 5. x = 4.5 is a good next trial.

2 Use trial and improvement to find the solution to the equation

$$x^3 - 5x = 60$$

Give your answer correct to 1 decimal place.

x	x³ − 5x	Comment
4	44	Too low
5	100	Too high
4.5	68.625	Too high
4.3	58.007	Too low
4.4	63.184	Too high
4.35	60.562…	Too high

(4 marks)

x = 4.3 (to 1 decimal place)

Keep trying different values. Make sure you write down the results of every trial.

x = 4.3 is too low and x = 4.4 is too high, so you know the answer is between 4.3 and 4.4. But you don't know which value is closer.

Try x = 4.35

x = 4.35 is too high, so you know the answer is between x = 4.3 and x = 4.35 This means it is closer to x = 4.3

Write down the answer correct to 1 decimal place.

Now try this

1 Use trial and improvement to find the solution to $2x^2 + 5x = 29$
Give your answer correct to 1 decimal place. *(4 marks)*

x	2x² + 5x	Comment
2	18	Too low

2 Use trial and improvement to find the solution to $x(x + 3)(x - 1) = 54$
Give your answer correct to 1 decimal place. *(4 marks)*

x	x(x + 3)(x − 1)	Comment
3	36	Too low

Remember $2x^2$ means $2 \times x^2$

61

A* A B C D

The quadratic formula

This is how the quadratic formula will appear on the formula sheet in your exam.

The quadratic formula

The solutions of $ax^2 + bx + c = 0$ where $a \neq 0$, are given by

$$x = \frac{-b \pm \sqrt{(b^2 - 4ac)}}{2a}$$

You may need to use this when answering more demanding problem-solving questions.

Safe substituting

Equation is in the form $ax^2 + bx + c = 0$.

Write down your values of *a*, *b* and *c* before you substitute.

Use brackets when you are substituting negative numbers.

Show what you have substituted in the formula.

Simplify what is under the square root and write this down.

The ± symbol means you need to do two calculations.

Worked example

target A

Solve $5x^2 + x + 11 = 14$

Give your solutions correct to 3 significant figures.   *(3 marks)*

$5x^2 + x - 3 = 0$

$a = 5, b = 1, c = -3$

$$x = \frac{-1 \pm \sqrt{1^2 - 4 \times 5 \times (-3)}}{2 \times 5}$$

$$= \frac{-1 + \sqrt{61}}{10} \text{ or } \frac{-1 - \sqrt{61}}{10}$$

$= 0.681024...$ or $-0.881024...$

$= 0.681$ or -0.881 (to 3 s.f.)

You are asked to find 'solutions'. This tells you that you are solving a quadratic equation.

You must give your answer 'correct to 3 significant figures'. This tells you that you need to use the quadratic formula. Turn to the formula sheet.

Write down at least five figures after the decimal point on the calculator display before giving your final answer. You might need to use the S⇔D button on your calculator to get your answer as a decimal.

How many solutions?

A quadratic equation can have two solutions, one solution or no solutions.

You can use $b^2 - 4ac$ (the part under the square root) to work out how many solutions a quadratic equation has.

You can't calculate the square root of a negative number.

If $b^2 - 4ac$ is negative, there are no solutions.

If $b^2 - 4ac = 0$, there is only one solution. — ±0 appears in the formula, so you get the same answer whether you use + or −.

If $b^2 - 4ac > 0$, there are two different solutions.

Now try this

target A

1 Solve $2x^2 + 3x - 8 = 0$
Give your answers to 2 decimal places.
You **must** show your working.   *(3 marks)*

2 Solve $3x^2 - 2x - 10 = 0$
Give your answers to 3 significant figures.
You **must** show your working.   *(3 marks)*

$$x = \frac{-b \pm \sqrt{b^2 - 4ac}}{2a}$$

The division line goes under **all** the terms in the numerator.

You have to divide the whole of $-b \pm \sqrt{b^2 - 4ac}$ by $2a$

Quadratic graphs

An equation which contains an x^2 term is called a QUADRATIC equation. Quadratic equations have CURVED graphs. You can draw the graph of a quadratic equation by completing a table of values.

Worked example

target C

(a) Complete the table of values for $y = 4x - x^2$ *(2 marks)*

x	−1	0	1	2	3	4	5
y	−5	0	3	4	3	0	−5

(b) On the grid, draw the graph of $y = 4x - x^2$ *(2 marks)*

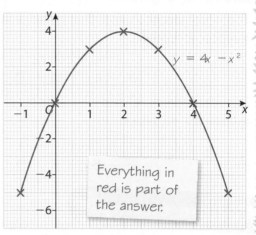

$y = 4x - x^2$

Everything in red is part of the answer.

Substitute each value of x into the equation to get a corresponding value of y.

When $x = -1$: $4 \times -1 - (-1)^2 = -4 - 1 = -5$

When $x = 4$: $4 \times 4 - 4^2 = 16 - 16 = 0$

Plot your points carefully on the graph and join them with a **smooth** curve.

Check it!
All the points on your graph should lie on the curve. If one of the points doesn't fit then double-check your calculation.

Drawing quadratic curves

Use a sharp pencil. ✓

Plot the points carefully. ✓

Draw a smooth curve that passes through every point. ✓

Label your graph. ✓

Shape of graph will be either ∪ or ∩

Drawing a smooth curve

It's easier to draw a smooth curve if you turn your graph paper so your hand is INSIDE the curve.

Now try this

target C

(a) Complete the table of values for $y = x^2 - 2x - 4$.

x	−2	−1	0	1	2	3	4
y		−1	−4			−1	4

(2 marks)

(b) On the grid, draw the graph of $y = x^2 - 2x - 4$. *(2 marks)*

(c) On the grid, draw the graph of $y = 1$. *(1 mark)*

(d) Write down the coordinates of the points of intersection of the two graphs. *(2 marks)*

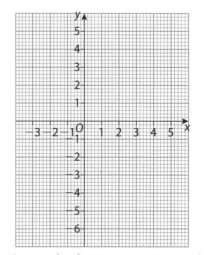

A* A B C D

Using quadratic graphs

You can use graphs to solve quadratic equations. You will need to look for the x-values of the points where a quadratic graph intersects with a straight line.

This grid shows one quadratic graph and two straight-line graphs.

The x-values at A and B are the solutions of the quadratic equation

$$x^2 - 4x = 12$$
or $$x^2 - 4x - 12 = 0$$

The x-values at C and D are the solutions of the quadratic equation

$$x^2 - 4x = x + 3$$
or $$x^2 - 5x - 3 = 0$$

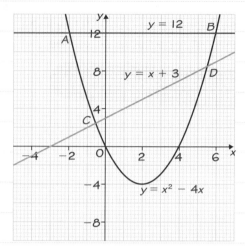

Worked example

target B

This is a graph of $y = 2x^2 + 5x$

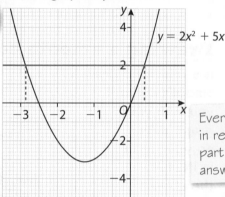

$y = 2x^2 + 5x$

Everything in red is part of the answer.

By drawing a <u>suitable straight line on the graph</u>, solve the equation $2x^2 + 5x - 2 = 0$

Give your answers correct to 1 decimal place.

(3 marks)

$$2x^2 + 5x - 2 = 0 \qquad (+2)$$
$$2x^2 + 5x = 2$$
$$x = 0.4, x = -2.9$$

You can solve the **quadratic equation** $2x^2 + 5x - 2 = 0$ by finding where the graph $y = 2x^2 + 5x$ crosses the straight line $y = 2$.

Draw the line $y = 2$ on the graph.

The solutions are the x-values at the points of intersection.

An appropriate line

To find an APPROPRIATE straight line, rearrange the quadratic equation so that the left-hand side matches the equation of the quadratic graph. The right-hand side will tell you which line to draw. Here is another example:

Graph given:	$y = x^2 - 6x + 4$
Equation to solve:	$x^2 - 5x + 1 = 0$
Rearrange equation:	$x^2 - 6x + 4 = -x + 3$
Line to draw:	$y = -x + 3$

Now try this

target C-A

(a) Complete the table of values for $y = x^2 + 3x - 3$.

x	−5	−4	−3	−2	−1	0	1	2
y	7	1			−3	1	7	

(2 marks)

(b) On a grid with $-5 \leqslant x \leqslant 5$ and $-6 \leqslant y \leqslant 6$, draw the graph of $y = x^2 + 3x - 3$.   (2 marks)

(c) Write down the solutions of $x^2 + 3x - 3 = 0$.   (1 mark)

(d) By drawing a suitable straight line, work out the solutions of the equation $x^2 + 2x - 4 = 0$.   (3 marks)

Graphs of curves

You need to know the general shapes of these three types of graph, and what their equations look like.

 Graphs of the form $y = \dfrac{k}{x}$ are called RECIPROCAL GRAPHS.

  Graphs of the form $y = a^x$ or $y = a^{-x}$ are called EXPONENTIAL GRAPHS.

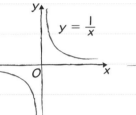 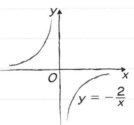

$y = \dfrac{1}{x}$ $y = -\dfrac{2}{x}$

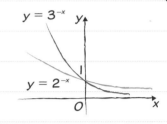

$y = 2^x$ $y = 3^{-x}$ $y = \left(\dfrac{1}{2}\right)^x$ $y = 2^{-x}$

 Graphs which contain an x^3 term are called CUBIC GRAPHS.

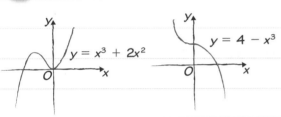

$y = x^3 + 2x^2$ $y = 4 - x^3$

Exponential growth

Exponential graphs appear in equations representing growth or decay. If there are 500 bacteria in a Petri dish, and the number doubles every hour, then after h hours there will be $N = 500 \times 2^h$ bacteria in the Petri dish. The graph of N against h will be an exponential graph.

You can use the $\boxed{x^\square}$ key to work out powers larger than 3 on your calculator.

Worked example

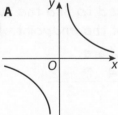 A   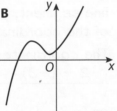 B   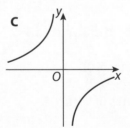 C   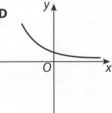 D

Write down the letter of the graph which could have the equation

(a) $y = 0.5^{-x}$ .....D.....   (b) $y = \dfrac{2}{x}$ .....A.....   (c) $y = x^3 + 3x^2 + 2x + 1$ .....B.....

(1 mark)   (1 mark)   (1 mark)

Now try this

Two points (6, 1) and (1, 6) on the graph of $y = \dfrac{6}{x}$ for $x > 0$ are plotted.

(a) Complete a sketch of the graph of $y = \dfrac{6}{x}$ for $x > 0$
(2 marks)

(b) Work out the exact coordinates of the point where this curve intersects the line $y = x$
(2 marks)

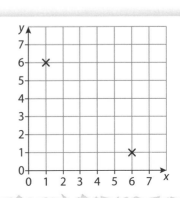

A* A B C D

3-D coordinates

You can describe a point in three dimensions by using coordinates that have three numbers.

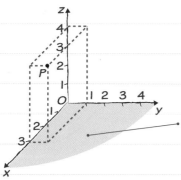

This is a normal coordinate grid with x- and y-axes.

You can add a z-axis which is perpendicular to both the other axes.

You always give coordinates in the order (x, y, z).

The coordinates of point P are (3, 1, 4).

The arrow on each axis points in the positive direction.

Finding missing coordinates

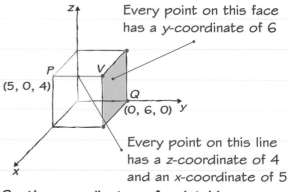

Every point on this face has a y-coordinate of 6

Every point on this line has a z-coordinate of 4 and an x-coordinate of 5

So the coordinates of point V are (5, 6, 4).

Worked example

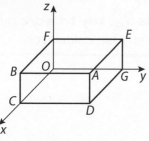

The diagram shows a cuboid drawn on a 3-D grid.
Vertex D has coordinates (5, 6, 0).
Vertex B has coordinates (5, 0, 2).
Write down the coordinates of vertex A.

(2 marks)

(5, 6, 2)

Line segments

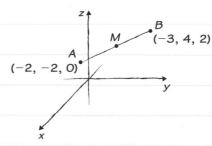

To find the coordinates of the midpoint of a line segment, you need to find the average of the coordinates of the endpoints.

The coordinates of M are

$$\left(\frac{-2 + (-3)}{2}, \frac{-2 + 4}{2}, \frac{0 + 2}{2}\right)$$

or $(-2\frac{1}{2}, 1, 1)$.

For a reminder about line segments in two dimensions look at page 40.

Now try this

The diagram shows a cuboid drawn on a 3-D grid.
B is the point (4, 9, 0)
G is the point (0, 9, 5)
Work out the coordinates of:

(a) A   **(b)** D   **(c)** F   *(3 marks)*
(d) M, the mid-point of EF   *(2 marks)*
(e) N, the mid-point of EC   *(2 marks)*

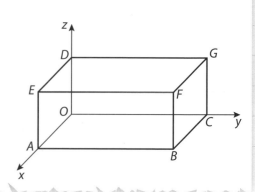

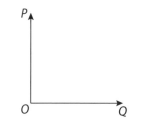

Proportionality formulae

You can answer some tricky proportionality questions quickly by remembering the proportionality FORMULAE and the shapes of the proportionality GRAPHS.

Proportionality in words	Using $\propto$	Formula
y is directly proportional to x	$y \propto x$	$y = kx$
y is directly proportional to the square of x	$y \propto x^2$	$y = kx^2$
y is directly proportional to the cube of x	$y \propto x^3$	$y = kx^3$
y is directly proportional to the square root of x	$y \propto \sqrt{x}$	$y = k\sqrt{x}$
y is inversely proportional to x	$y \propto \dfrac{1}{x}$	$y = \dfrac{k}{x}$
y is inversely proportional to the square of x	$y \propto \dfrac{1}{x^2}$	$y = \dfrac{k}{x^2}$

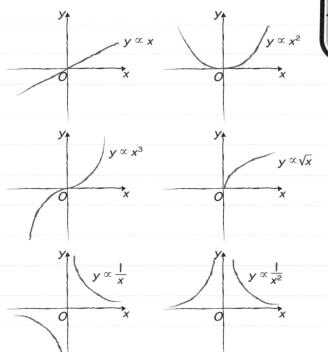

$y \propto x$ $y \propto x^2$ $y \propto x^3$ $y \propto \sqrt{x}$ $y \propto \dfrac{1}{x}$ $y \propto \dfrac{1}{x^2}$

Worked example

target A

q is inversely proportional to the square of t.
When $t = 4$, $q = 8.5$
Calculate the value of q when $t = 5$. (5 marks)

$q \propto \dfrac{1}{t^2}$

$q = \dfrac{k}{t^2}$

$8.5 = \dfrac{k}{4^2}$

$k = 8.5 \times 4^2 = 136$

$q = \dfrac{136}{t^2}$

When $t = 5$: $q = \dfrac{136}{5^2} = 5.44$

Always...

1. Write down the statement of proportionality and then the formula.
2. Substitute the values you are given.
3. Solve the equation to find k.
4. Write down the formula using the value of k.
5. Use your formula to find any unknown values.

◀ Don't round your answer unless the question tells you to.

Now try this

target A

(a) P is inversely proportional to Q.
When $P = 75$, $Q = 96$
Express P in terms of Q. (3 marks)

(b) P and Q are positive quantities.
Sketch a graph of the relationship between P and Q on the diagram on the right. (1 mark)

(c) Calculate the value of Q when P is twice as big as Q. (2 marks)

P
O Q

A*
A
B
C
D

Transformations 1

You can change the equation of a graph to translate it, stretch it or reflect it. These tables show you how you can use functions to transform the graph of $y = f(x)$.

Function	$y = f(x) + a$	$y = f(x + a)$	$y = af(x)$
Transformation of graph	Translation $\begin{pmatrix} 0 \\ a \end{pmatrix}$	Translation $\begin{pmatrix} -a \\ 0 \end{pmatrix}$	Stretch in the vertical direction, scale factor a
Useful to know	$f(x) + a \rightarrow$ move UP a units $f(x) - a \rightarrow$ move DOWN a units	$f(x + a) \rightarrow$ move LEFT a units $f(x - a) \rightarrow$ move RIGHT a units	x-values stay the same
Example	$y = f(x) + 3$ $y = f(x)$	$y = f(x)$ $y = f(x + 5)$	$y = 3f(x)$ $y = f(x)$

Function	$y = f(ax)$	$y = -f(x)$	$y = f(-x)$
Transformation of graph	Stretch in the horizontal direction, scale factor $\frac{1}{a}$	Reflection in the x-axis	Reflection in the y-axis
Useful to know	y-values stay the same	'−' outside the bracket	'−' inside the bracket
Example	$y = f(2x)$ $y = f(x)$	$y = f(x)$ $y = -f(x)$	$y = f(-x)$ $y = f(x)$

Worked example

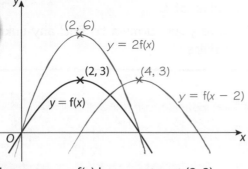

The curve $y = f(x)$ has a vertex at $(2, 3)$.

Write down the coordinates of the vertex of the curve with equation

(a) $y = f(x - 2)$ $(4, 3)$ *(1 mark)*

(b) $y = 2f(x)$ $(2, 6)$ *(1 mark)*

$y = f(x - 2)$ is a translation 2 units right along the x-axis.

$y = 2f(x)$ is a stretch in the vertical direction, scale factor 2.

Now try this

This is the graph of $y = x^2 - 2$

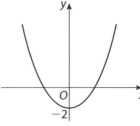

(a) The graph is translated by vector $\begin{pmatrix} 0 \\ 6 \end{pmatrix}$.
Draw the position of the graph after this translation. *(1 mark)*

(b) The new graph is then translated by vector $\begin{pmatrix} 3 \\ 0 \end{pmatrix}$. Draw the position of the graph after this second translation. *(1 mark)*

(c) Work out the equation of the graph after both translations. Give your answer in the form $y = ax^2 + bx + c$. *(3 marks)*

Transformations 2

You need to be able to convert between function notation and equations of graphs.
This table shows some transformations that may come up in your exam.

Original function	$y = 2x + 3$	$y = \sin x°$	$y = x^2 - 2x + 1$	$y = x^2$
Transformation	$f(x) \to f(x) + 2$	$f(x) \to f(x - 30)$	$f(x) \to 2f(x)$	$f(x) \to f(3x)$
Which means...	movement UP by 2 units	movement RIGHT by 30°	stretch in vertical direction, scale factor 2	stretch in horizontal direction, scale factor $\frac{1}{3}$
New function	$y = 2x + 5$	$y = \sin(x - 30)°$	$y = 2x^2 - 4x + 2$	$y = 9x^2$

Graphs of sine and cosine

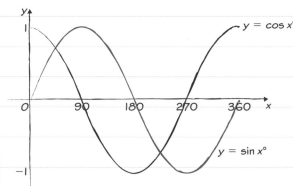

The graph of $y = \cos x°$ is identical to the graph of $y = \sin x°$ except that it has been moved to the left by 90°.

Write down the transformations using function notation.

(a) Stretch in the vertical direction with scale factor $\frac{1}{2}$.

(b) Stretch in the horizontal direction with scale factor 2.

Worked example

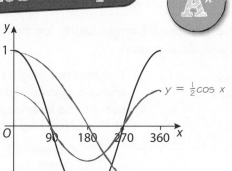

The graph of $y = \cos x$ is shown for $0° \leq x \leq 360°$.

On the same diagram, draw the following graphs for $0° \leq x \leq 360°$.

(a) $y = \frac{1}{2}\cos x$ (1 mark)

$y = \frac{1}{2}f(x)$

(b) $y = \cos\left(\frac{1}{2}x\right)$ (1 mark)

$y = f\left(\frac{1}{2}x\right)$

Now try this

The diagram shows the graph of $y = \sin x$ for $0° \leq x \leq 360°$.

On the diagram, draw each of these graphs, and write down each transformation using function notation.

(a) $y = 2\sin x$ (2 marks)

(b) $y = \sin 2x$ (2 marks)

(c) $y = \sin(x - 90°)$ (2 marks)

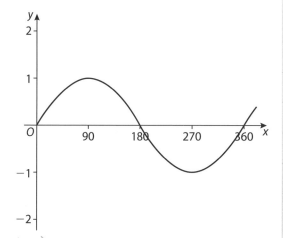

69

Angle properties

You need to remember all of these angle properties and their correct names.

CORRESPONDING ANGLES are equal.

ALTERNATE ANGLES are equal.

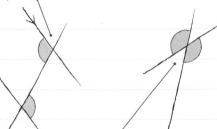

ALLIED ANGLES (or INTERIOR ANGLES) add up to 180°.

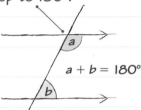

$a + b = 180°$

Parallel lines are marked with arrows.

OPPOSITE ANGLES are equal.

These are useful angle facts for triangles and parallelograms:

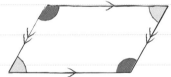

Interior angle

The exterior angle of a triangle is equal to the sum of the interior angles at the other two vertices.

Exterior angle

The opposite angles of a parallelogram are equal.

You need to know the proofs of the angle properties of triangles and quadrilaterals.

Golden rule

When answering angle problems, you need to give a reason for each step of your working.

Geometric proof checklist

To prove a geometric fact you need to:
• write down each step of your working clearly ✓
• give a reason for each step of your working ✓
• use the correct words for your reasons. ✓

Worked example

target **D**

This diagram shows a triangle and a straight line. *PQ* is parallel to *RS*.

Prove that the angles in the triangle add up to 180°. You can use the fact that the angles on a straight line add up to 180°. *(4 marks)*

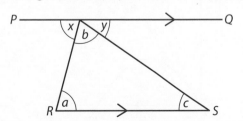

$a = x$ (alternate angles are equal)
$c = y$ (alternate angles are equal)
$x + b + y = 180°$ (angles on a straight line add up to 180°)
So $a + b + c = 180°$.

EXAM ALERT!

You need to remember this proof for your exam.

Students have struggled with exam questions similar to this – **be prepared**

Now try this

target **C**

AB is parallel to *CD*.

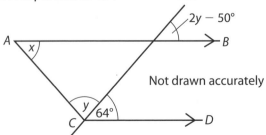

$2y - 50°$

Not drawn accurately

Work out the value of *x*. *(5 marks)*

Solving angle problems

You might need to use angle properties to solve problems in your exam. Remember to give reasons for every step of your working.

Reasons

Use these reasons in angle problems:
- Angles on a straight line add up to 180°.
- Angles around a point add up to 360°.
- Opposite angles are equal.
- Corresponding angles are equal.
- Alternate angles are equal.
- Angles in a triangle add up to 180°.
- Angles in a quadrilateral add up to 360°.
- Base angles of an isosceles triangle are equal.

Use the properties on the diagram:
AB is parallel to CD
AC is parallel to BD
BE is equal to DE

Worked example　target D

Work out the size of the angle marked x. Give reasons for each step of your working.

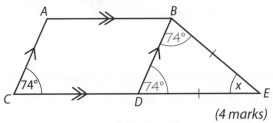

(4 marks)

∠BDE = 74° (corresponding angles are equal)

∠DBE = 74° (base angles in an isosceles triangle are equal)

x + 74° + 74° = 180° (angles in a triangle add up to 180°)

x = 180° − 148°

x = 32°

Worked example　target C

ABCD is a quadrilateral.
Angle A is a right angle.
Angles B and C are in the ratio 1 : 2.
Angle D is 70° more than angle B.

Work out the size of angles B, C and D.

(4 marks)

Write angle B as x then express angles C and D in terms of x. Angle problems could involve ratio or proportion, or you might need to write your own equation. For a reminder about writing your own equations to solve problems look at page 44.

A = 90°
C = 2x
D = x + 70°
A + B + C + D = 360° (angles in a quadrilateral add up to 360°)
So 90° + x + 2x + (x + 70°) = 360°
4x = 200°
x = 50°

So B = 50°, C = 2 × 50° = 100° and D = 50° + 70° = 120°

Now try this　target C

Triangle ABC is isosceles and triangle BDC is isosceles.

Angle DBC = 2x.

Show clearly that angle BAC = 2x.

(4 marks)

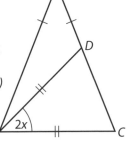

A* A A B C D

Angles in polygons

Polygon questions are all about interior and exterior angles.

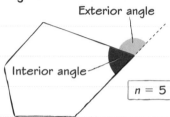

Exterior angle

Interior angle

$n = 5$

Use these formulae for a polygon with n sides.

Sum of interior angles = $180° \times (n - 2)$

Sum of exterior angles = $360°$

This diagram shows part of a REGULAR polygon with 30 sides.

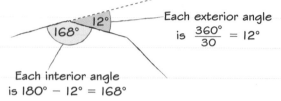

168°   12°

Each exterior angle is $\frac{360°}{30} = 12°$

Each interior angle is $180° - 12° = 168°$

Don't try to draw a 30-sided polygon! If there's no diagram given in a polygon question, you probably don't need to draw one.

Regular polygons

In a regular polygon all the sides are equal and all the angles are equal.

If a regular polygon has n sides then each exterior angle is $\frac{360°}{n}$

Regular pentagon $\frac{360°}{5}$ 72°

Regular hexagon   60° $\frac{360°}{6}$

Regular octagon $\frac{360°}{8}$ 45°

You can use the fact that the angles on a straight line add up to 180° to work out the size of one of the interior angles.

Worked example

target **C**

The diagram shows part of a regular polygon.

Work out the number of sides in the polygon.

156°

(3 marks)

Exterior angle = $180° - 156° = 24°$

$\frac{360°}{n}$ so $n = \frac{360°}{24°} = 15$

The polygon has 15 sides.

EXAM ALERT!

It's usually easier to work with **exterior angles** than interior angles. Use the fact that angles on a straight line add up to 180° to calculate the size of one exterior angle.

In a regular n-sided polygon, each exterior angle is $\frac{360°}{n}$. You can use this fact to write an equation and solve it to find n.

Students have struggled with exam questions similar to this – **be prepared!**

Now try this

target **C**

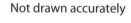

Not drawn accurately

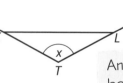

JK, KL and LM are three sides of a 20-sided regular polygon.

JK and ML are extended to meet at T.

Work out the size of angle KTL, marked x on the diagram.   (4 marks)

Angles TKL and TLK are both external angles of the 20-sided polygon.

Circle facts

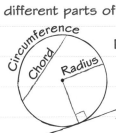

You need to know the names of the different parts of a circle.

Diameter = radius × 2

Tangent

The other parts of a circle are shown on pages 99 and 100.

When you are solving circle problems:

- correctly identify the angle to be found
- use all the information given in the question
- mark all calculated angles on the diagram
- give a reason for each step of your working.

You might need to use angle facts about triangles, quadrilaterals and parallel lines in circle questions. There is a list of angle facts on page 71.

A*
A
B
C
D

Key circle facts

 1 The angle between a radius and a tangent is 90°.

 2 Two tangents which meet at a point outside a circle are the same length.

3 A triangle which has one vertex at the centre of a circle and two vertices on the circumference is an ISOSCELES TRIANGLE.

Each short side of the triangle is a radius, so they are the same length.

Remember that the base angles of an isosceles triangle are equal.

Worked example

A *A and B are points on the circumference of a circle centre O. AC and BC are both tangents to the circle. Angle BCA = 42°. Work out the size of the angle marked x.*   *(3 marks)*

AC = BC (tangents from a point outside a circle are the same length)

$\angle ABC = \dfrac{180° - 42°}{2} = 69°$

(base angles in an isosceles △ are equal, and angles in a △ add up to 180°)

$x + 69° = 90°$ (angle between a tangent and a radius = 90°)

$x = 21°$

AC = BC, so mark these lines with a dash. Make sure you write down the circle fact you are using as well. To write a really good answer you have to give a reason for each step of your working.

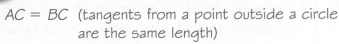

Now try this

 A *TQP and TSR are tangents to the circle, centre O, radius 7 cm.*

TQ = 16 cm.

W is the point where TO meets the circumference.

Work out the distance TW.   *(4 marks)*

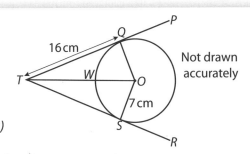

Not drawn accurately

A*
A
B
C
D

Circle theorems

If you're aiming for an A or A* you need to LEARN these six circle theorems.

1 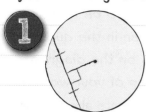 The perpendicular from a chord to the centre of the circle bisects the chord.

2  The angle at the centre of the circle is twice the angle on the circumference.

3  The angle in a semicircle is 90°.

4  Angles in the same segment are equal.

5  Opposite angles of a cyclic quadrilateral add up to 180°.

6 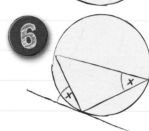 The angle between a tangent and a chord is equal to the angle in the alternate segment.

This is called the ALTERNATE SEGMENT THEOREM.

See page 73 for more circle facts.

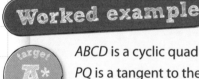

target **A***

ABCD is a cyclic quadrilateral.
PQ is a tangent to the circle at A.

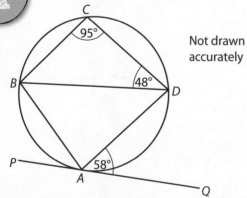

Not drawn accurately

Prove that BC is parallel to AD.   (5 marks)

∠ABD = 58° (alternate segment theorem)

∠DAB = 85° (opposite angles in cyclic quadrilateral add up to 180°)

∠BDA = 180° − (58° + 85°) = 37° (angles in a triangle add up to 180°)

∠DBC = 180° − (95° + 48°) = 37° (angles in a triangle add up to 180°)

So ∠BDA = ∠DBC are alternate angles, so BC is parallel to AD.

Prove it's parallel

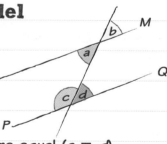

To PROVE that LM is parallel to PQ you need to show that ONE of these is true:

ALTERNATE ANGLES are equal (a = d)

CORRESPONDING ANGLES are equal (b = d)

ALLIED ANGLES add up to 180° (a + c = 180°)

target **A***

ABCD is a cyclic quadrilateral.
PBQ is a tangent to the circle at B.

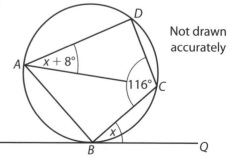

Not drawn accurately

Work out the value of x.   (4 marks)

Perimeter and area

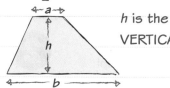

A*
A
B
C
D

Triangle

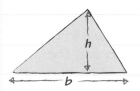

Area $= \frac{1}{2}bh$
Learn this formula ✓

Parallelogram

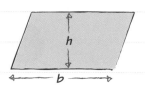

Area $= bh$
Learn this formula ✓

Trapezium

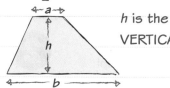

h is the
VERTICAL HEIGHT

Area $= \frac{1}{2}(a + b)h$
Given on the formula sheet ✓

You can calculate areas and perimeters of more complex shapes by splitting them into parts.

You might need to draw some extra lines on your diagram and add or subtract areas.

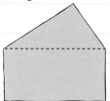

Area = rectangle + triangle   Area = triangle − rectangle

Area basics

Lengths are all in the same units. ✓

Give units with the answer. ✓

Lengths in cm means area units are cm². ✓

No calculator. ✓

Lengths in m means area units are m². ✓

Worked example

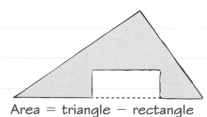

The diagram shows a garden bed.

Adrian wants to cover the bed with grass seed.

A packet of grass seed will cover 10 m².

(a) How many packets of grass seed does Adrian need to buy?   *(3 marks)*

Area = 6 × 4.5 + 4 × 3 = 27 + 12 = 39 m²

Adrian needs to buy 4 packets of grass seed.

Adrian also wants to build a fence around the edge of the garden bed.

(b) Calculate the total length of Adrian's fence.   *(1 mark)*

6 + 4.5 + 3 + 4 + 3 + 8.5 = 29 m

Draw a dotted line to divide the diagram into two rectangles.

You have to use the information in the question to work out the missing lengths. The diagram is **not accurately drawn**, so you can't use a ruler to measure.

8.5 m − 4.5 m = 4 m

6 m − 3 m = 3 m

Write these lengths on your diagram.

Make sure you answer the question that has been asked.

You need to say how many packets of grass seed Adrian needs to buy.

Now try this

This is a sketch of a house.

(a) Work out the area of this shape. State the units of your answer.   *(4 marks)*

(b) Work out the perimeter of this shape.   *(1 mark)*

The shape is made from a rectangle and a trapezium.

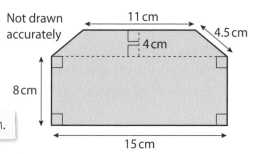

Not drawn accurately

75

Similar shapes 1

Shapes are SIMILAR if one shape is an enlargement of the other.

SIMILAR TRIANGLES satisfy one of these three conditions:

 1 All three pairs of angles are equal.

 2 All three pairs of sides are in the same ratio.

 3 Two sides are in the same ratio and the included angle is equal.

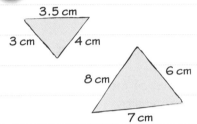

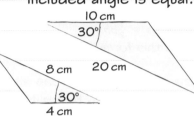

Worked example

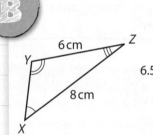

XYZ and ABC are similar triangles.

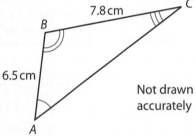

Not drawn accurately

(a) Work out the length of AC. *(3 marks)*

$$\frac{AC}{XZ} = \frac{BC}{YZ}$$
$$\frac{AC}{8} = \frac{7.8}{6}$$
$$AC = \frac{7.8 \times 8}{6}$$
$$= 10.4 \text{ cm}$$

(b) Work out the length of XY. *(3 marks)*

$$\frac{XY}{AB} = \frac{YZ}{BC}$$
$$\frac{XY}{6.5} = \frac{6}{7.8}$$
$$XY = \frac{6 \times 6.5}{7.8}$$
$$= 5 \text{ cm}$$

Start with the unknown length on top of a fraction. Make sure you write your ratios in the correct order.

Similar shapes checklist

Use these facts to solve similar shapes problems:

Corresponding angles equal. ✓

Corresponding sides in same ratio. ✓

Spotting similar triangles

Here are some similar triangles:

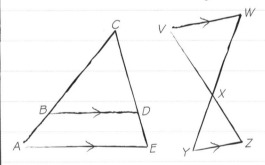

Triangle ACE is similar to triangle BCD.

Triangle VWX is similar to triangle ZYX.

Now try this

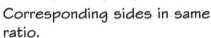

Triangles ABC and PQR are similar.

Angle ACB = angle PRQ.

(a) Work out the size of angle PRQ. *(2 marks)*

(b) Work out the length of PQ. *(2 marks)*

Not drawn accurately

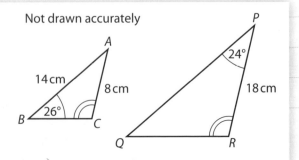

Congruent triangle proof

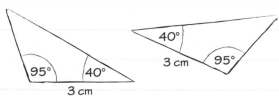

Two triangles are CONGRUENT if they have exactly the same shape and size.
To prove this you have to show that ONE of these four conditions is true.

1   **SSS** (three sides are equal)

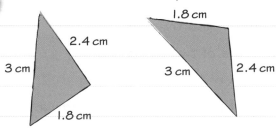

2   **AAS** (two angles and a corresponding side are equal)

3   **SAS** (two sides and the included angle are equal)

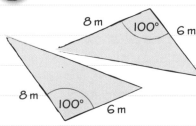

The angle must be BETWEEN the two sides for SAS.

4   **RHS** (right angle, hypotenuse and a side are equal)

Worked example

ABCD and *DEFG* are squares.

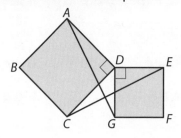

Prove that triangles *AGD* and *ECD* are congruent.   *(4 marks)*

AD = *DC* (two sides of same square)
DE = *DG* (two sides of same square)
∠*CDE* = ∠*CDG* + 90°
∠*GDA* = ∠*CDG* + 90°
So ∠*CDE* = ∠*GDA*
So *AGD* is congruent to *ECD* (SAS)

EXAM ALERT!

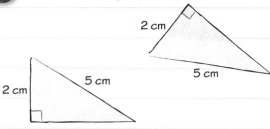

You need to show that one of the conditions for congruence is true. You can only use the properties given in the question to prove congruence.

Always write down the condition for congruence.

> Students have struggled with exam questions similar to this – **be prepared!**

Common sides

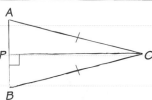

If two triangles have a side in COMMON then those two sides are equal.

PC is common to triangles *APC* and *BPC*. Both triangles have a right angle and the same hypotenuse, so they satisfy RHS and are congruent.

Now try this

ABC is an isosceles triangle, in which *AB* = *AC*.
H and *K* are points on *AB* and *AC* such that *AH* = *AK*.
Prove that triangles *BHC* and *CKB* are congruent.
State the reason for congruency.   *(4 marks)*

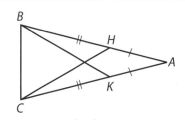

A*
A
B
C
D

Pythagoras' theorem

Pythagoras' theorem is a really useful rule. You can use it to find the length of a missing side in a right-angled triangle.

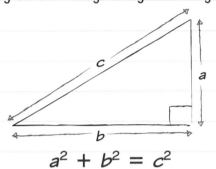

$$a^2 + b^2 = c^2$$

Pythagoras checklist

short^2 + short^2 = long^2 ✓

Right-angled triangle.   ✓

Lengths of two sides known.   ✓

Length of third side missing.   ✓

Learn this. It's not on the formula sheet.   ✓

Worked example

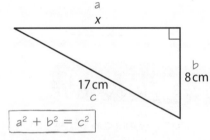

target C

This right-angled triangle has sides x, 17 cm and 8 cm.

$$a^2 + b^2 = c^2$$

Show that $x = 15$ cm   *(3 marks)*

$$x^2 + 8^2 = 17^2$$
$$x^2 = 17^2 - 8^2$$
$$= 225$$
$$x = \sqrt{225} = 15 \text{ cm}$$

The question says 'show that' so you have to show **all** your working. Be careful when the missing length is one of the **shorter** sides.
1. Label the longest side of the triangle c.
2. Label the other two sides a and b.
3. Write out the formula for Pythagoras' theorem.
4. Substitute the values for a, b and c into the formula.
5. Rearrange the formula and solve. Make sure you show **every step** in your working.
6. Write units in your answer.

Pythagoras questions come in lots of different forms. Just look for the right-angled triangle.

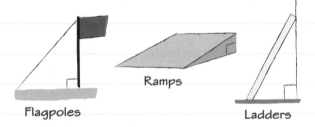

Flagpoles   Ramps   Ladders

Calculator skills

Use these buttons to find squares and square roots with your calculator.

 x^2 $\sqrt{\square}$

You might need to use the S⇔D key to get your answer as a decimal number.

Now try this

target C

(a) Work out the value of y.   *(3 marks)*

(b) Use your value of y to work out the value of z.   *(3 marks)*

12 cm

Not drawn accurately

z

y

12.5 cm

8.4 cm

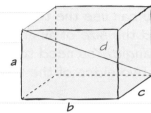

Pythagoras in 3-D

To tackle the most demanding questions, you need to be able to use Pythagoras' theorem in 3-D shapes.

You can use Pythagoras' theorem to find the length of the longest diagonal in a cuboid.

You can also use Pythagoras to find missing lengths in pyramids and cones.

$$a^2 + b^2 + c^2 = d^2$$

Why does it work?

You can use 2-D Pythagoras twice to show why the formula for 3-D Pythagoras works.

$$x^2 = b^2 + c^2$$

$$d^2 = a^2 + x^2$$
$$= a^2 + b^2 + c^2$$

Worked example

The diagram shows a cuboid. Work out the length of PQ.
(3 marks)

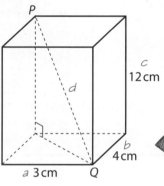

$$d^2 = a^2 + b^2 + c^2$$
$$PQ^2 = 3^2 + 4^2 + 12^2$$
$$= 169$$
$$PQ = \sqrt{169} = 13$$
So PQ is 13 cm.

Write out the formula for Pythagoras in 3-D. Label the sides of the cuboid a, b and c, and label the long diagonal d.

You could also answer this question by sketching two right-angled triangles and using 2-D Pythagoras.

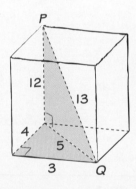

Check it!

The diagonal must be longer than any of the other three lengths.

13 cm looks about right. ✓

Now try this

This is a diagram of a wedge.

Angles TXW, TXY and XYW are all 90°.

(a) Work out the length of TW.

Give your answer to 1 decimal place. *(3 marks)*

(b) Without further calculation, give reasons as to which of angles TWX or TYX is larger. *(2 marks)*

Not drawn accurately

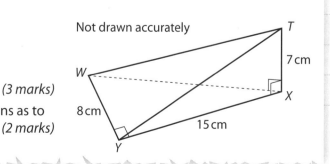

A*
A
B
C
D

Trigonometry 1

You can use the trigonometric ratios to find the size of an angle in a right-angled triangle. You need to know the lengths of two sides of the triangle.

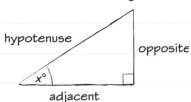

hypotenuse

opposite

$x°$

adjacent

The sides of the triangle are labelled relative to the ANGLE you need to find.

Trigonometric ratios

$\sin x° = \dfrac{\text{opp}}{\text{hyp}}$ (remember this as S^O_H)

$\cos x° = \dfrac{\text{adj}}{\text{hyp}}$ (remember this as C^A_H)

$\tan x° = \dfrac{\text{opp}}{\text{adj}}$ (remember this as T^O_A)

You can use $S^O_H C^A_H T^O_A$ to remember these rules for trig ratios.

These rules only work for RIGHT-ANGLED triangles.

Worked example

target **B**

Calculate the size of angle x. *(3 marks)*

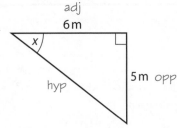

adj
6 m

x

5 m opp

hyp

$\tan x° = \dfrac{\text{opp}}{\text{adj}} = \dfrac{5}{6}$

$x° = 39.805\,571\,09° = 39.8°$ (to 3 s.f.)

Label the **hyp**otenuse first — it's the longest side.

Then label the side **adj**acent to the angle you want to work out.

Finally label the side **opp**osite the angle you want to work out.

Remember $S^O_H C^A_H T^O_A$. You know **opp** and **adj** here so use T^O_A.

Do **not** 'divide by tan' to get x on its own. You need to use the $\tan^{-1}$ function on your calculator.

$\tan^{-1}\left(\dfrac{5}{6}\right)$

39.80557109

Write down all the figures on your calculator display then round your answer.

Using your calculator

To find a missing angle using trigonometry you have to use one of these functions.

$\sin^{-1}$ $\cos^{-1}$ $\tan^{-1}$

These are called INVERSE TRIGONOMETRIC functions. They are the inverse operations of sin, cos and tan.

Make sure that your calculator is in **degree mode**. Look for the **D** symbol at the top of the display.

Now try this

target **B**

Work out the size of angle x in each of these triangles. Give your answers correct to 1 decimal place.

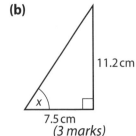

(a)

4.3 cm   6.1 cm

x

(3 marks)

(b)

11.2 cm

x

7.5 cm
(3 marks)

(c)

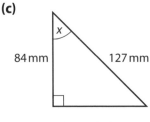

x

84 mm   127 mm

(3 marks)

Trigonometry 2

You can use the trigonometric ratios to find the length of a missing side in a right-angled triangle. You need to know the length of another side and the size of one of the acute angles.

Worked example

Calculate the length of side a. *(3 marks)*

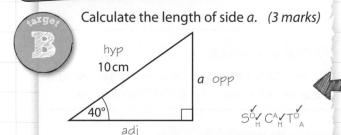

hyp
10 cm

a opp

40°

adj

$S^{\checkmark}_H \ C^A_H{}^{\checkmark} \ T^O_A{}^{\checkmark}$

$\sin x° = \dfrac{opp}{hyp}$

$\sin 40° = \dfrac{a}{10}$

$a = 10 \times \sin 40°$

$\quad = 6.42787\ldots$

$\quad = 6.43 \text{ cm (to 3 s.f.)}$

Label the sides of the triangle relative to the 40° angle. Write $S^O_H C^A_H T^O_A$ and tick the pieces of information you have. You need to use S^O_H here.

Write the values you know in the rule and replace **opp** with a. You can solve this equation to find the value of a.

Write down at least four figures of the calculator display before giving your final answer correct to 3 significant figures.

Check it!

Side a must be shorter than the hypotenuse. 6.43 cm looks about right. ✓

Angles of elevation and depression

Some trigonometry questions will involve angles of elevation and depression.

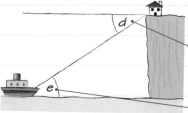

The angle of depression of the ship from the house.

The angle of elevation of the house from the ship.

Angles of elevation and depression are always measured from the horizontal.

In this diagram, d = e because they are alternate angles.

Now try this

Work out the length of side a in each of these triangles. Give your answers correct to 1 decimal place.

In part **(c)** a is the hypotenuse. It will be on the bottom of the fraction when you substitute, so be careful with your calculation.

(a)

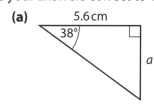

5.6 cm
38°

a

(3 marks)

(b)

a
73°

18.5 cm

(3 marks)

(c)

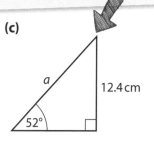

a

12.4 cm

52°

(3 marks)

The sine rule

The SINE RULE applies to any triangle. You don't need a right angle.

You label the angles of the triangle with capital letters and the sides with lower case letters. Each side has the same letter as its OPPOSITE angle.

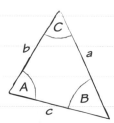

$$\frac{a}{\sin A} = \frac{b}{\sin B} = \frac{c}{\sin C}$$

This version is given on the formula sheet. Use it to find a missing side.

$$\frac{\sin A}{a} = \frac{\sin B}{b} = \frac{\sin C}{c}$$

Learn this version. It's useful for finding a missing angle.

Worked example

 target A

Calculate the size of angle x. *(3 marks)*

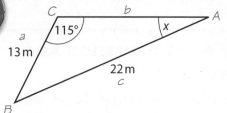

$$\frac{\sin A}{a} = \frac{\sin C}{c}$$

$$\frac{\sin x}{13} = \frac{\sin 115°}{22}$$

$$\sin x = \frac{13 \times \sin 115°}{22}$$

$$= 0.5355...$$

$$x = 32.4° \text{ (3 s.f.)}$$

Golden rule

To use the sine rule you need to know a side length and the OPPOSITE angle.

EXAM ALERT!

This is not a right-angled triangle so you can't use $S^O{}_H C^A{}_H T^O{}_A$. To find an angle use the 'upside down' version of the sine rule. You're not interested in side *b* or angle *B* so ignore this part of the rule.

Start by writing the value you want to find on top of the first fraction. Then substitute the other values you know and solve an equation to find *x*. Use the $\sin^{-1}$ function on your calculator.

Students have struggled with exam questions similar to this – **be prepared!**

Worked example

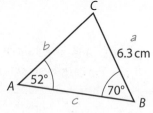 target A

Show that AC = 7.5 cm. *(3 marks)*

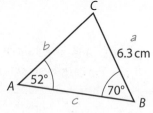

$$\frac{b}{\sin B} = \frac{a}{\sin A}$$

$$\frac{AC}{\sin 70°} = \frac{6.3}{\sin 52°}$$

$$AC = \frac{6.3 \times \sin 70°}{\sin 52°}$$

$$= 7.5126...$$

$$= 7.5 \text{ cm (2 s.f.)}$$

You know a side length and the opposite angle so you can use the sine rule.

Check it!
The greater side length is opposite the greater angle. ✓

Now try this

target A

In triangle ABC, AB = 11 cm, AC = 13 cm and angle ABC = 57°.

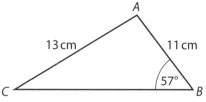

(a) Work out the size of angle ACB. *(3 marks)*
(b) Work out the length of BC. *(3 marks)*

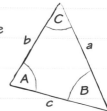

The cosine rule

The COSINE RULE applies to any triangle. You don't need a right angle.

You usually use the cosine rule when you are given two sides and the included angle (SAS) or when you are given three sides and want to work out an angle (SSS).

$$a^2 = b^2 + c^2 - 2bc \cos A$$

This version is on the formula sheet. Use it to find a missing side.

$$\cos A = \frac{b^2 + c^2 - a^2}{2bc}$$

Learn this version. It's useful for finding a missing angle.

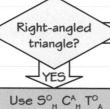

Which rule?

This chart shows you which rule to use when solving trigonometry problems in triangles:

Right-angled triangle? → NO → Side and the opposite angle given? → NO → Use the cosine rule

↓ YES

Use $S^O_H\, C^A_H\, T^O_A$

↓ YES

Use the sine rule

Worked example

 target A

PQRS is a trapezium. Work out the length of the diagonal PR. *(3 marks)*

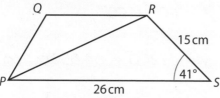

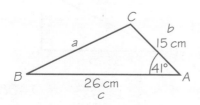

$a^2 = b^2 + c^2 - 2bc \cos A$

$PR^2 = 15^2 + 26^2 - 2 \times 15 \times 26 \times \cos 41°$

$\quad = 312.3265...$

$PR = 17.6727... = 17.7 \text{ cm (3 s.f.)}$

If you are given a more complicated diagram it is sometimes useful to sketch a triangle. Label your triangle with a as the missing side.

This is not a right-angled triangle so you can't use $S^O_H C^A_H T^O_A$. You know two sides and the included angle (SAS) so you can use the cosine rule.

Substitute the values you know into the formula. Work out the right-hand side using your calculator, but don't round your answer yet.

Use √☐ Ans on your calculator to find the final answer.

Round to 3 significant figures or to the same degree of accuracy as the original measurements. Because this answer shows all its workings, you could give either 17.7 cm or 18 cm as the anwer.

Now try this

 target A

Triangle PQR has sides of 9 cm, 10 cm and 14 cm.

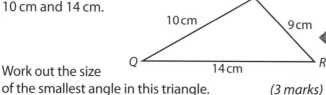

Work out the size of the smallest angle in this triangle. *(3 marks)*

Use the cosine rule if you are given **three sides** and you need to find an angle. You should use this version of the cosine rule:

$$\cos A = \frac{b^2 + c^2 - a^2}{2bc}$$

Remember to check that your final answer makes sense.

A*
A
B
C
D

Trigonometry in 3-D

You can use $S^O_H C^A_H T^O_A$ to find the angle between a LINE and a PLANE.

You might need to combine trigonometry and Pythagoras' theorem when you are solving 3-D problems.

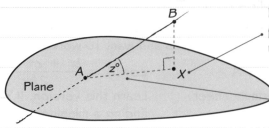

Point X is directly below point B, so ABX is a right-angled triangle.

Angle z is the angle between the line and the plane.

target
A*

The diagram shows a triangular prism.

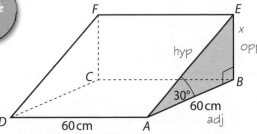

Calculate the angle between the line DE and the base of the prism.   *(6 marks)*

△ABE: you know one angle and the adjacent side. You are looking for the opposite side, so use T^O_A.

△DAB: you know two sides so you can use Pythagoras' theorem.

△DEB: you know the opposite and adjacent sides so use T^O_A. Use $\tan^{-1}$ to find the value of z.

Do **not** round any of your answers until the end — write down at least six figures from each calculator display.

$\tan 30° = \dfrac{x}{60}$

$x = 60 × \tan 30°$

$ = 34.6410... \text{ cm}$

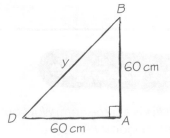

$y^2 = 60^2 + 60^2$

$ = 7200$

$y = \sqrt{7200} = 84.8528... \text{ cm}$

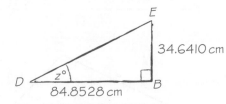

$\tan z° = \dfrac{34.6410...}{84.8528...} = 0.408\,24...$

$z° = \tan^{-1} 0.408\,24... = 22.2076...$

$ = 22.2° \text{ (to 3 s.f.)}$

target
A*

$ML = 28$ cm and $JK = 40$ cm

Angle $MKL = 32°$

Angles MLK, MLJ and JKL are all 90°.

Work out the size of angle MJL.   *(6 marks)*

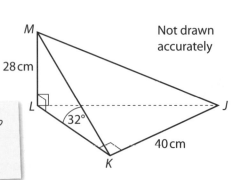

Not drawn accurately

Draw sketches of triangles MLK, LJK and MJL to keep track of your working. Remember you might need to combine trigonometry and Pythagoras' theorem.

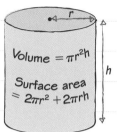

A*
A
B
C
D

Circles and cylinders

You need to learn these formulae for circles and cylinders. They're not on the formula sheet.

Circle

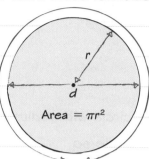

Circumference $= 2\pi r$
$= \pi d$

Area $= \pi r^2$

Cylinder

$\text{Volume} = \pi r^2 h$

$\text{Surface area} = 2\pi r^2 + 2\pi r h$

Worked example

target C

The diagram shows a game counter in the shape of a semicircle.

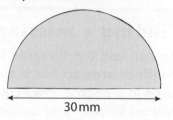

30 mm

Work out the area of the counter.　　*(3 marks)*

Radius $= 30 \div 2 = 15$ mm
Area of circle $= \pi r^2$
$\qquad = \pi \times 15^2$
$\qquad = 706.8583...$ mm^2
Area of counter $= 706.8583... \div 2$
$\qquad = 350$ mm^2 (2 s.f.)

Calculator skills

Make sure you know how to enter π on your calculator. On some calculators you have to press these keys:

SHIFT　　π　e
[×10^x]　➡　[×10^x]

EXAM ALERT!

The formula for the area of a circle uses the **radius**. If the length shown on the diagram is the **diameter**, you need to divide it by 2 before you substitute into the formula. Don't round any values until the end of your working.

> Students have struggled with exam questions similar to this – **be prepared!**

In terms of π

Unless a question asks you for a specific degree of accuracy, you can give your answers as a whole number or fraction multiplied by π. An answer given in terms of π is an EXACT ANSWER rather than a ROUNDED ANSWER.

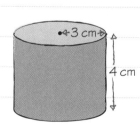
3 cm
4 cm

Volume of cylinder $= \pi r^2 h$
$\qquad = \pi \times 3^2 \times 4$
EXACT ANSWER
Volume $= 36\pi$ cm^3
ROUNDED ANSWER
Volume $= 113$ cm^3 (to 3 s.f.)

Now try this

target C

This shape is made from a rectangle and a semicircle.
Work out the area of the shape.
Give your answer to a sensible degree of accuracy.　　*(4 marks)*

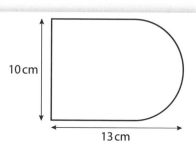

10 cm

13 cm

A*
A
B
C
D

Sectors of circles

Each pair of radii divides a circle into two sectors, a MAJOR SECTOR and a MINOR SECTOR.

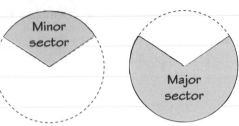

Minor sector

Major sector

You can find the area of a sector by working out what fraction it is of the whole circle.

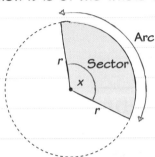

Arc

Sector

r

x

r

For a sector with angle x of a circle with radius r:

Sector = $\frac{x}{360°}$ of the whole circle so

Area of sector = $\frac{x}{360°} \times \pi r^2$

Arc length = $\frac{x}{360°} \times 2\pi r$

Learn these formulae. ✓

You can give answers in terms of π. ✓

Worked example

target **A**

The diagram shows a minor sector of a circle of radius 13 cm.

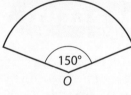

150°

O

Work out the perimeter of the sector.   *(4 marks)*

Arc length = $\frac{x}{360°} \times 2\pi r$

$= \frac{150°}{360°} \times 2\pi \times 13$

$= 34.03392...$

Perimeter = arc length + radius + radius

$= 34.03392... + 13 + 13$

$= 60$ cm (2 s.f.)

Don't round until your final answer. The radius is given correct to 2 significant figures so this is a good degree of accuracy.

Finding a missing angle

You can use the formulae for arc length or area to find a missing angle in a sector. Practise this method to help you tackle the hardest questions.

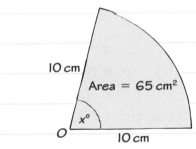

10 cm

Area = 65 cm²

$x°$

O

10 cm

Area of sector = $\frac{x}{360} \times \pi r^2$

$65 = \frac{x}{360} \times \pi (10)^2$

$x = \frac{65 \times 360}{\pi (10)^2}$

$= 74.4845...$

$= 74.5°$ (to 3 s.f.)

Now try this

target **A***

AB is the arc length of a minor sector of a circle, centre O, radius 12 cm.

AB measures 27 cm.

Work out the size of angle AOB, marked x on the diagram. Give your answer correct to 3 significant figures.   *(3 marks)*

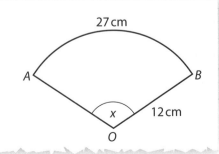

27 cm

A B

x 12 cm

O

Triangles and segments

When you know the lengths of two sides and the angle BETWEEN THEM, the area of any triangle can be found using this formula.

Area = $\frac{1}{2}$ ab sin C

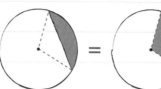

You can use this formula for ANY triangle. You don't need to have a right angle.

This formula is on the formula sheet.

Areas of segments

A chord divides a circle into two SEGMENTS.

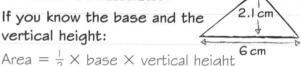

| Area of segment | = | Area of whole sector | − | Area of triangle |

The diagram shows a sector of a circle with centre O.

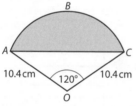

Work out the area of the shaded segment ABC. Give your answer correct to 3 significant figures. **(5 marks)**

Whole sector OABC:
Area = $\frac{120}{360}$ × π × 10.4²
    = 113.2648... cm²

Triangle OAC:
Area = $\frac{1}{2}$ × 10.4 × 10.4 × sin 120°
    = 46.8346... cm²

Shaded segment ABC:
Area = 113.2648... − 46.8346...
    = 66.4302...
    = 66.4 cm² (to 3 s.f)

This is a very common exam question! You need to be able to calculate the area of a sector and a triangle.

To get full marks you need to keep track of your working. Make sure you write down exactly what you are calculating at each step.

Remember that 10.4 cm is the length of one side of the triangle **and** the radius of the circle.

Make sure you don't round too soon. Write down all the figures from your calculator display at each step. Only round your **final answer** to 3 significant figures.

Which formula?

If you know the base and the vertical height:

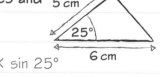

Area = $\frac{1}{2}$ × base × vertical height
    = $\frac{1}{2}$ × 6 × 2.1
    = 6.3 cm²

If you know two sides and the included angle:

Area = $\frac{1}{2}$ ab sin C
    = $\frac{1}{2}$ × 5 × 6 × sin 25°
    = 6.3... cm²

Here is a circle, radius 4 cm, with angle AOC = 135°.

Using the fact that sin 135° = sin 45° = $\frac{1}{\sqrt{2}}$, show clearly that the area of the minor segment ABC = $6\pi - 4\sqrt{2}$ cm². **(5 marks)**

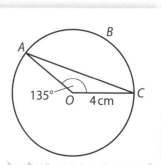

A*
A
B
C
D

Prisms

A prism is a 3-D shape with a constant CROSS-SECTION. If the cross-section is a rectangle then we call the prism a cuboid.

Cuboid

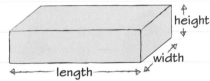

↑ height ↓
width
←――― length ―――→

Volume = length × width × height   Learn this formula ✓

Prism

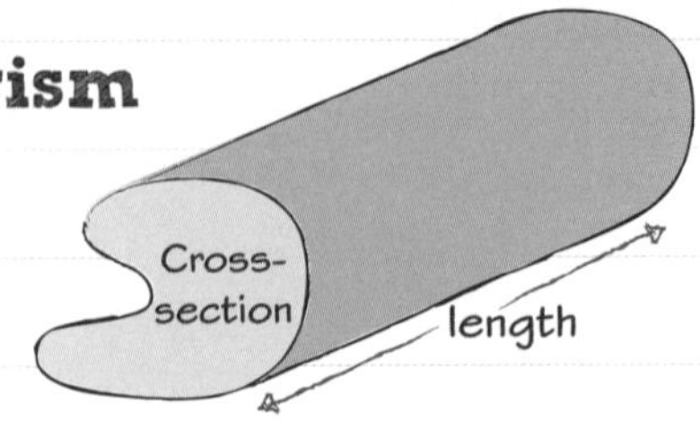

Cross-section
length

Volume = area of cross-section × length

Given on the formula sheet ✓

You might need to work out the area of the cross-section before working out the volume of the prism.

You should draw a sketch of the cross-section separately to work out its area.

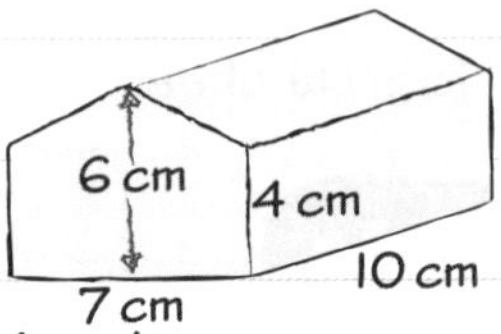

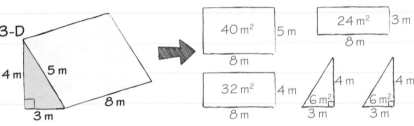

6 cm   4 cm
7 cm   10 cm

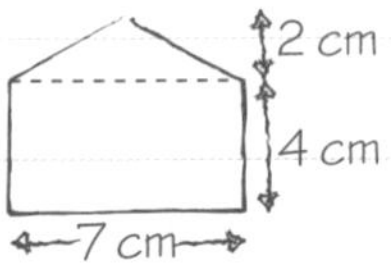

↑2 cm
↓
4 cm
←―7 cm―→

Area of cross-section = area of rectangle + area of triangle

$$= 7 \times 4 + \frac{1}{2} \times 7 \times 2 = 35\,\text{cm}^2$$

Volume of prism = 35 × 10 = 350 cm³

Surface area

To work out the surface area of a 3-D shape, you need to add together the areas of all the faces.

It's a good idea to sketch each face with its dimensions.

Remember to include the faces that you can't see.

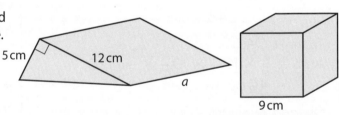

4 m   5 m
3 m   8 m

40 m²   5 m
8 m

24 m²   3 m
8 m

32 m²   4 m
8 m

4 m
6 m²
3 m

4 m
6 m²
3 m

Surface area = 40 + 32 + 24 + 6 + 6 = 108 m²

Worked example

target C

The diagram shows a triangular prism and a cube. They both have the **same** volume.

Work out the length of *a*.   *(4 marks)*

5 cm   12 cm
a

9 cm

Volume of cube= 9³ = 729 cm³

Volume of prism = Area of cross-
section × length

$$= \frac{1}{2} \times 5 \times 12 \times a$$

$$= 30a$$

30a = 729

a = 24.3 cm

> Calculate the volume of the cube, and write an expression for the volume of the prism. Set these equal to each other and solve the equation to find *a*.

Now try this

target C

The diagram shows a 3-D shape with a cross-section made from a rectangle and a triangle.

Calculate the volume of the shape.

State the units of your answer.   *(5 marks)*

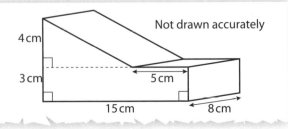

Not drawn accurately

4 cm
3 cm
15 cm
5 cm
8 cm

Volumes of 3-D shapes

A*
A
A
B
C
D

Cylinder

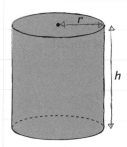

Volume of cylinder

= area of base × height

= $\pi r^2 h$

Pyramid

Volume of pyramid

= $\frac{1}{3}$ × area of base

× vertical height

= $\frac{1}{3}Ah$

Learn these volume formulae ✓

Cone

Volume of cone

= $\frac{1}{3}$ × area of base × vertical height

= $\frac{1}{3}\pi r^2 h$

Sphere

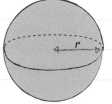

Volume of sphere = $\frac{4}{3}\pi r^3$

These volume formulae are on the formula sheet ✓

There is more about density on page 97.

Worked example

target **A**

The diagram shows a solid cone. It is made from aluminium with a density of 2.7 g/cm³.

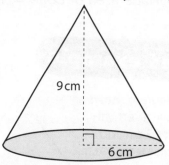

9 cm

6 cm

Work out the mass of the cone.   *(3 marks)*

Volume = $\frac{1}{3}\pi r^2 h$

$= \frac{1}{3} \times \pi \times 6^2 \times 9$

$= 108\pi \, cm^3$

Mass = $108\pi \times 2.7 = 916\,g$ (3 s.f.)

Now try this

target **B**

1 These two cylinders have the same volume.

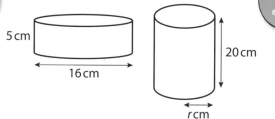

5 cm

16 cm

20 cm

r cm

Work out the radius, r of the second cylinder.   *(4 marks)*

Give your fraction in the lowest terms possible.

target **A***

2 A tennis ball of diameter 7 cm is packaged in a cylindrical box. The ball touches the sides, top and base of the box.

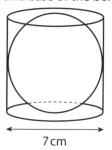

7 cm

Work out the fraction of the volume of the box taken up by the tennis ball. You must show your working.   *(5 marks)*

A* A B C D

Surface area

Cone

The formula for the CURVED SURFACE AREA of a cone is given on the formula sheet.

Curved surface area of cone = $\pi r \ell$

Be careful! This formula uses the slant height, ℓ, of the cone.

To calculate the TOTAL surface area of the cone you need to add the area of the base. Surface area of cone = $\pi r^2 + \pi r \ell$

Sphere

The formula for the surface area of a sphere is given on the formula sheet.

Surface area of sphere = $4\pi r^2$

For a reminder about areas of circles and surface areas of cylinders have a look at page 85.

A hemisphere is half a sphere, so the area of the curved surface is $\frac{1}{2} \times 4\pi r^2$.

Worked example

target **A**

The diagram shows a cone with vertical height 12 cm and base diameter 10 cm.

Work out the curved surface area of the cone.

(4 marks)

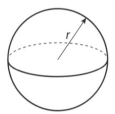

12 cm

10 cm

$r = 5$

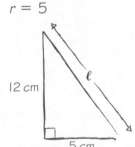

12 cm

ℓ

5 cm

$l^2 = 12^2 + 5^2 = 169$

$l = 13$ cm

Curved surface area
$= \pi r l$
$= \pi \times 13 \times 5$
$= 65\pi$ cm^2

To work out the curved surface area you need to know the radius and the slant height. You are given the **diameter** and the **vertical height**.

The radius is half the diameter = 5 cm.

To calculate the slant height you need to use Pythagoras' theorem. Sketch the right-angled triangle containing the missing length.

You can leave your answer in terms of π.

Compound shapes

You can calculate the surface area of more complicated shapes by adding together the surface area of each part.

4 cm

6 cm

Surface area $= \pi(4)^2 + 2\pi(4)(6) + \frac{1}{2}[4\pi(4)^2]$
$= 96\pi$ cm^2

Now try this

target **A**

A cone has a base diameter of 16 cm and a vertical height of 15 cm.

The cone is cut in half vertically through the vertex.

The diagram shows one of the half-cones.

Work out the total surface area of the half-cone.

Give your answer correct to 3 significant figures.   *(5 marks)*

15 cm

16 cm

A*
A
B
C
D

Plans and elevations

Plans and elevations are 2-D drawings of 3-D shapes as seen from different directions.

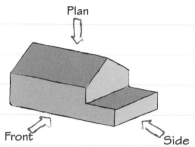
Plan
Front
Side

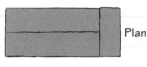
Plan

The PLAN is the view from above.

Front elevation

The FRONT ELEVATION is the view from the front.

Side elevation

This line shows a change in depth.

The SIDE ELEVATION is the view from the side.

You might be asked to draw a 3-D shape on isometric paper.

From above, 4 cubes can be seen.

Plan

From the side, 5 cubes can be seen.

From the front, 4 cubes can be seen.

Front elevation   Side elevation

There are a total of 7 cubes in this shape.

Front   Side

Worked example

target **D**

The diagram shows a solid shape.

On the grid below draw a plan, front and side elevations of the shape.
(3 marks)

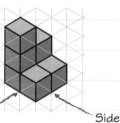
Front   Side

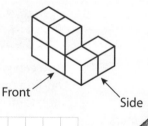
Plan   Front elevation   Side elevation

Imagine tracing an image of the shape on each side of a box.

Unfold the box to get your plan and elevations.

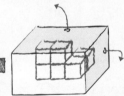

Put lines within the plan and side elevation to show where there is a change in height or depth.

Now try this

target **D-C**

This solid is made from centimetre cubes.

On separate 1 cm grids draw:

(a) the plan view of the solid   (1 mark)

(b) the front elevation of the solid   (1 mark)

(c) the side elevation of the solid.   (1 mark)

(d) Work out the total surface area of the solid.
(2 marks)

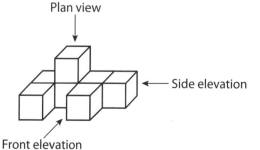
Plan view
Side elevation
Front elevation

91

A*
A
B
C
D

Bearings

Bearings are measured CLOCKWISE from NORTH.

Bearings always have three FIGURES. You might need to add zeros if the angle is less than 100°. For instance, in this diagram the bearing of B from A is 048°.

You can measure a bearing bigger than 180° by measuring this angle and subtracting it from 360°.

The bearing of C from A is 360° − 109° = 251°

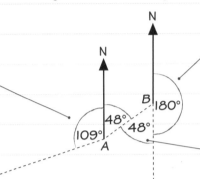

You can work out a reverse bearing by adding or subtracting 180°.

The bearing of A from B is 180° + 048° = 228°

These are alternate angles.

Worked example

 target **D**

Three paths meet at O. B is due East of O.

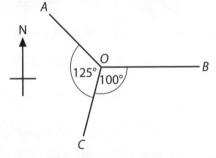

(a) Jake walks from O to A.
What bearing does he walk on?   *(1 mark)*

90° + 100° + 125° = 315°

(b) Delvinder walks from C to O.
What bearing does she walk on?   *(2 marks)*

Bearing of C from O = 90° + 100°
            = 190°

Bearing of O from C = 190° − 180° = 010°

Compass points

You need to know the compass points:

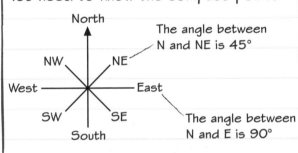

The angle between N and NE is 45°

The angle between N and E is 90°

EXAM ALERT!

You can work out a **reverse** bearing by adding or subtracting 180°. If the angle is greater than 180° then subtract. Remember to write your final answer as a **three-figure bearing.** You need to write 010°, not just 10°.

Students have struggled with exam questions similar to this – **be prepared!**

Now try this

 target **D**

An aircraft flies from A to B. The grid shows the positions of A and B.

(a) Use the grid to work out the actual distance AB. *(1 mark)*

(b) Measure and write down the three figure bearing of B from A.   *(1 mark)*

(c) The aircraft then flies to C.
The bearing of C from A is 120°. The bearing of C from B is 080°. Mark the position of C on the grid.   *(3 marks)*

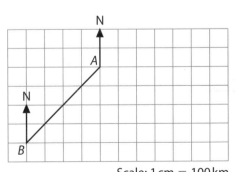

Scale: 1 cm = 100 km

A*
A
B
C
D

Scale drawings and maps

This is a SCALE DRAWING of the
Queen Mary II cruise ship.

Scale = 1 : 1000

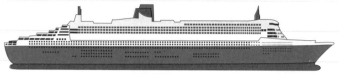

←————34.5 cm————→

You can use the scale to work out the
length of the actual ship.

34.5 × 1000 = 34 500

The ship is 34 500 cm or 345 m long.

Map scales

Map scales can be
written in different
ways:

- 1 to 25 000
- 1 cm represents 25 000 cm
- 1 cm represents 250 m
- 4 cm represent 1 km

MAP

SCALE
1 : 25 000

Worked example

The diagram shows a scale drawing
of a port and a lighthouse.

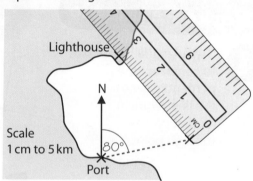

Lighthouse

N

Scale
1 cm to 5 km

80°

Port

A boat sails 12 km in a straight line on a
bearing of 080°.

(a) Mark the new position of the boat with a
cross.   *(2 marks)*

(b) How far away is the boat from the
lighthouse? Give your answer in km.
                                          (1 mark)

15 km

For a reminder about bearings have a
look at page 92.

Start by working out how far the boat is
from the port on the scale drawing.

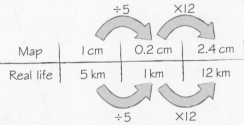

	÷5 →	×12 →	
Map	1 cm	0.2 cm	2.4 cm
Real life	5 km	1 km	12 km
	÷5 →	×12 →	

Now place the centre of your protractor
on the port with the zero line pointing
North. Put a dot at 80°. Line up your
ruler between the port and the dot. Draw
a cross 2.4 cm from the port.

Use a ruler to measure the distance from
the lighthouse to the boat. 3 cm on the
drawing represents 15 km in real life.

Now try this

(a) A map uses a scale of 1 : 150 000.
Two towns are 6 cm apart on the map.
How far apart are they in real life?
Give your answer in kilometres.   *(3 marks)*

(b) Two other towns are 15 miles apart in real
life.
How far apart will they be on the map?
                                          (3 marks)

5 miles = 8 km

A* A B C D

Constructions

'Construct' means 'draw accurately using a ruler and compasses'. You should make sure you have a good pair of compasses with stiff arms and a sharp pencil in your exam.

Worked example

1 Use ruler and compasses to **construct** a triangle with sides of length 3 cm, 4 cm and 5.5 cm.
(2 marks)

4 cm   3 cm
5.5 cm

Draw and label one side with a ruler. Then use your compasses to find the other vertex.

Worked example

2 Use ruler and compasses to **construct** the bisector of angle *PQR*.   *(2 marks)*

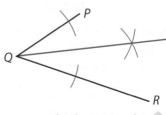
P
Q
R

Mark points on each arm an equal distance from *Q*. Then use arcs to find a third point an equal distance from these two points.

Worked example

3 Use ruler and compasses to **construct** the perpendicular bisector of the line *AB*.
(2 marks)

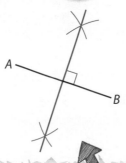
A
B

Use your compasses to draw intersecting arcs with centres at *A* and *B*. Remember to show **all** your construction lines and arcs.

Worked example

C

4 Use ruler and compasses to **construct** the perpendicular to the line segment *AB* that passes through point *P*.
(2 marks)

A   P   B

Use your compasses to mark two points on the line an equal distance from *P*. Then widen your compasses and draw arcs with their centres at these two points.

Worked example

5 Use ruler and compasses to **construct** the perpendicular to the line segment *AB* that passes through point *P*.   *(2 marks)*

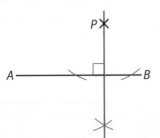
P
A   B

Use your compasses to mark two points on the line an equal distance from *P*. Draw two arcs with their centres at these points.

Now try this

(a) Construct an angle of 60° at *A*.

A ———————————— B
(2 marks)

(b) Mark a point, *P*, 4 cm from *A* on the line of your 60° angle. Construct the perpendicular from *P* to the line segment *AB*.   *(2 marks)*

A*
A
B
C
D

Loci

A LOCUS is a set of points which satisfy a condition. You can construct loci using ruler and compasses. A set of points can lie inside a REGION rather than on a line or curve.

The locus of points which are 7 cm from A is the circle, centre A.

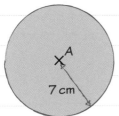

The region of points less than 7 cm from A lies inside this circle.

The locus of points which are the SAME DISTANCE from B as from C is the perpendicular bisector of BC.

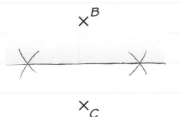

Points in the shaded region are closer to B than to C.

The locus of points which are 2 cm away from ST consists of two semicircles and two straight lines.

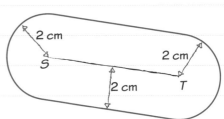

Combining conditions

You can be asked to shade a region which satisfies more than one condition.

Here, the shaded region is more than 6 cm from point D and closer to line BC than to line AD.

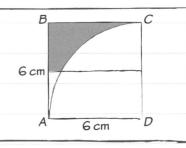

---

target **C**

The diagram shows part of a beach and the sea. 1 cm represents 20 m.

There is a lifeguard tower at point P.

Public swimming is allowed in a region of the sea less than 30 m from the lifeguard tower. Shade this region on the diagram.   *(2 marks)*

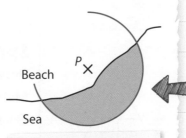

Beach   P ×

Sea

Everything in red is part of the answer.

1 cm represents 20 m so 1.5 cm represents 30 m.

There is more about scale drawing on page 93.

Set your compasses to 1.5 cm.

Set your compasses accurately by placing the point **on top** of your ruler at the 0 mark.

Remember to show all your construction lines.

target **C**

(a) Construct an accurate copy of this triangle.

(b) Shade the region of the triangle satisfied by these two conditions:

(i) Points must lie nearer to AB than BC.

(ii) Points must lie nearer to B than A.   *(5 marks)*

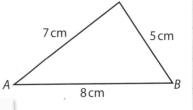

For part **(b)(i)** you need to construct the bisector of angle ABC.

A*
A
B
C
D

Speed

This is the formula triangle for speed.

Average speed • ← D → • Distance
• → Time
S T

Average speed = $\dfrac{\text{total distance travelled}}{\text{total time taken}}$

Time = $\dfrac{\text{distance}}{\text{average speed}}$

Distance = average speed × time

Using a formula triangle

Cover up the quantity you want to find with your finger.

The position of the other two quantities tells you the formula.

$T = \dfrac{D}{S}$ $S = \dfrac{D}{T}$ $D = S \times T$

Units

The most common units of speed are:

- metres per second: m/s
- kilometres per hour: km/h
- miles per hour: mph.

The units in your answer will depend on the units you use in the formula.

When distance is measured in km and time is measured in hours, speed will be measured in km/h.

When you are calculating a distance or time, you MUST make sure that the units of the other quantities match.

Minutes and hours

For questions on speed, you need to be able to convert between minutes and hours.

Remember there are 60 minutes in 1 hour.

To convert from minutes to hours you divide by 60.

24 minutes = 0.4 hours $\dfrac{24}{60} = \dfrac{2}{5} = 0.4$

To convert from hours to minutes you multiply by 60.

3.2 hours = 192 minutes   3.2 × 60 = 192
= 3 hours 12 minutes

Worked example

target **C**

A plane travels at a constant speed of 600 km/h for 45 minutes.

How far has it travelled? *(2 marks)*

D
S T

45 minutes = $\dfrac{45}{60}$ hours = $\dfrac{3}{4}$ hour

D = S × T
= $600 \times \dfrac{3}{4} = \dfrac{600 \times 3}{4} = \dfrac{1800}{4} = 450$

The plane has travelled 450 km.

Speed checklist

Draw formula triangle. ✓

Make sure units match. ✓

Give units with answer. ✓

Draw the formula triangle on your exam paper. You need to make sure the units match so start by converting 45 minutes into hours.

Now try this

target **D**

1 Bradley cycled 172 km at an average speed of 40 kilometres per hour.

How long did it take him?

Give your answer in hours and minutes.

(3 marks)

target **C**

2 Nisha drives from Newcastle to Oxford.

Her average speed is 55 mph.

The journey takes 4 hours 48 minutes

How far did she drive? *(3 marks)*

Density

The density of a material is its mass per unit volume.

This is the formula triangle for density.

$$\text{Density} = \frac{\text{mass}}{\text{volume}}$$

$$\text{Volume} = \frac{\text{mass}}{\text{density}}$$

$$\text{Mass} = \text{density} \times \text{volume}$$

Worked example

target C

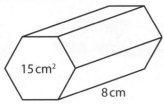

15 cm²

8 cm

The diagram shows a solid hexagonal prism.
The area of the cross-section of the prism is 15 cm².
The length of the prism is 8 cm.
The prism is made from wood with a density of 0.8 grams per cm³.
Work out the mass of the prism.   *(4 marks)*

Volume of prism
 = area of cross-section × length
 = 15 × 8
 = 120 cm³

$M = D \times V$
 = 0.8 × 120
 = 96

The mass of the prism is 96 g.

Units
The most common units of density are
- grams per cubic centimetre: g/cm³
- kilograms per cubic metre: kg/m³.

Mass = density × volume
You are given the density so you need to work out the volume of the prism.
The formula for the volume of a prism is given on the formula sheet.
The density is in grams per cm³ and the volume is in cm³ so the mass will be in grams.

Worked example

target C

An iron bar has a volume of 1.2 m³ and a mass of 9444 kg. Calculate the density of iron.
(2 marks)

$$D = \frac{M}{V} = \frac{9444}{1.2} = 7870 \text{ kg/m}^3$$

Volume is in m³ and mass is in kg so density will be in kg/m³.

Now try this

target C

1 This wedge is made from wood with a density of 0.65 g/cm³.

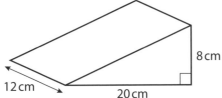

8 cm

12 cm   20 cm

Work out the mass of the wedge.   *(4 marks)*

2 A silver coin has a volume of 2.83 cm³ and a mass of 29.7 g.
Work out the density of silver.
Give your answer correct to 2 decimal places.
(2 marks)

Volume of prism = area of cross-section × length. For a reminder about volumes of prisms see page 88.

Converting units

You can convert between METRIC UNITS by multiplying or dividing by 10, 100 or 1000.

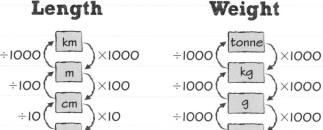

Length

÷1000 (km) ×1000
÷100 (m) ×100
÷10 (cm) ×10
(mm)

Weight

÷1000 (tonne) ×1000
÷1000 (kg) ×1000
÷1000 (g) ×1000
(mg)

Volume or capacity

÷1000
÷100 (litre) ×100
(cl) ×1000
÷10 (ml or cm³) ×10

1 m³ = 1000 litres

Converting compound units

To convert between measures of speed you need to convert one unit first then the other. Write the new units at each step of your working. To convert 72 km/h into m/s:

72 km/h → 72 × 1000 = 72 000 m/h

72 000 m/h → 72 000 ÷ 3600 = 20 m/s

1 hour = 60 × 60 = 3600 seconds

Imperial units

You need to remember these conversions for your exam.

Metric unit	Imperial unit
1 (kg)	2.2 pounds (lb)
1 litre (*l*)	$1\frac{3}{4}$ pints
4.5 litres	1 gallon
8 km	5 miles
30 cm	1 foot (ft)

When converting between imperial units you will be GIVEN the conversions.

Worked example

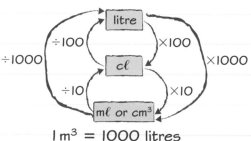

Worked example   target D

Heather drives along a motorway at an average speed of 60 mph.

Work out how long it takes her to drive 120 km.   *(3 marks)*

8 km = 5 miles

60 ÷ 5 = 12

8 × 12 = 96

60 mph = 96 km/h

$\text{Time} = \dfrac{\text{distance}}{\text{average speed}}$

$= \dfrac{120}{96} = 1\frac{1}{4}$ hours

First of all, you need to convert the speed from mph to km/h.

Convert 60 miles into km. You can use equivalent ratios:

miles : km
5 : 8
×12   ×12
60 : 96

Kilometres are smaller than miles so the speed in km/h should be a larger number than the speed in mph.

Now try this

1 In France, the speed limit on motorways, when it is dry, is 130 km/h. What speed is this in mph?   *(2 marks)*

2 Ian's car has a petrol tank that holds 12 gallons. He fills the tank, from empty, with petrol that costs £1.32 per litre. How much should he pay for this petrol?   *(3 marks)*

3 A cheetah can run at speeds up to 31 metres per second. Convert 31 m/s into a speed in km/h.   *(2 marks)*

A*
A
B
C
D

Translations, reflections and rotations

You might have to describe these transformations in your exam. To describe a translation you need to give a vector. To describe a reflection you need to give the equation of the mirror line. To describe a rotation you need to give the direction, the angle and the centre of rotation.

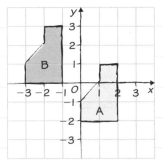

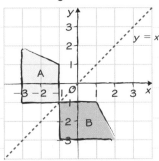

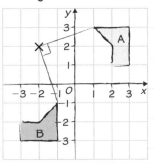

You can ask for tracing paper in an exam. This makes it easy to rotate shapes and check your answers.

A to B: TRANSLATION by the vector $\begin{pmatrix} -3 \\ 2 \end{pmatrix}$.

A to B: REFLECTION in the line $y = x$.

A to B: ROTATION 90° clockwise about the point (−2, 2).

For all three transformations, lengths of sides do not change, angles in shapes do not change and B is congruent to A.

Worked example

target **C**

The diagram shows two shapes P and Q.

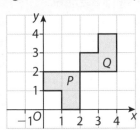

Describe fully the single transformation which takes shape P to shape Q. *(3 marks)*

Rotation 90° clockwise with centre (3, 1).

EXAM ALERT!

To **fully describe** a rotation you need to write down:
- the word 'rotation'
- the angle of turn and the direction
- the centre of rotation.

Be careful when the shapes are joined at one corner. This point is not necessarily the centre of rotation.

Check it!
You are allowed to ask for tracing paper in your exam. Trace shape P and put your pencil on your centre of rotation. Rotate the tracing paper to see if the shapes match up. ✓

Students have struggled with exam questions similar to this – **be prepared!**

Now try this

target **C**

Triangles A, B and C are shown on the grid.

(a) Describe fully the **single** transformation that takes triangle A onto triangle B. *(3 marks)*

(b) Write down the vector that describes the translation of triangle A onto triangle C. *(2 marks)*

The top number in your vector describes the horizontal translation and the bottom number describes the vertical translation.

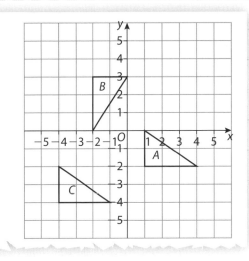

A*
A
B
C
D

Enlargements

To describe an enlargement you need to give the scale factor and the centre of enlargement.

The SCALE FACTOR of an enlargement tells you how much each length is multiplied by.

$$\text{Scale factor} = \frac{\text{enlarged length}}{\text{original length}}$$

Lines drawn through corresponding points on the object (A) and image (B) meet at the CENTRE OF ENLARGEMENT.

When the scale factor is between 0 and 1, image B is SMALLER than object A.

When the scale factor is negative, image B is on the OTHER SIDE of the centre of enlargement.

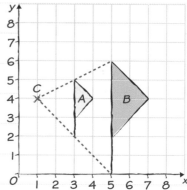

A to B: Each point on B is twice as far from C as the corresponding point on A.

Enlargement with scale factor 2, centre (1, 4).

For enlargements, angles in shapes do not change but lengths of sides do change.

Worked example

Triangle A is shown on the grid.

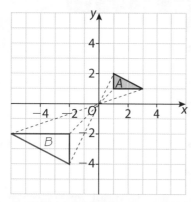

Enlarge triangle A by scale factor −2 with centre of enlargement O.   (3 marks)

The scale factor is **negative** so the image will be on the **opposite** side of the centre of enlargement and will be **upside down**.
Follow these steps to enlarge the triangle:

1. Draw lines from each vertex through the centre of enlargement.

2. Measure the distance from each point on triangle A to the centre of enlargement. The corresponding point on triangle B will be 2 times the distance from the centre of enlargement.

Check it!
Each length on triangle B should be 2 times the corresponding length on triangle A. ✓

Now try this

Triangle A is shown on the grid.

(a) Enlarge triangle A by scale factor $\frac{1}{2}$ with centre of enlargement (−3, 1).

Label the image B.   (2 marks)

(b) Enlarge triangle A by scale factor $-1\frac{1}{2}$ with centre of enlargement (0, 1).

Label the image C.   (3 marks)

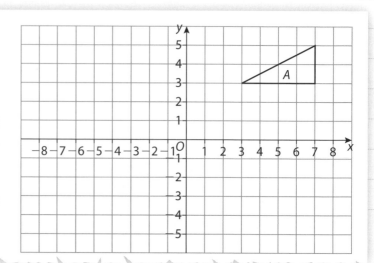

A*
A
B
C
D

Combining transformations

You can describe two or more transformations using a single transformation.

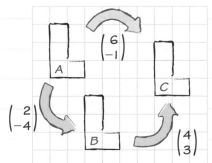

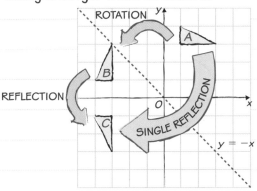

A to B to C: A translation $\begin{pmatrix} 2 \\ -4 \end{pmatrix}$ followed by a translation $\begin{pmatrix} 4 \\ 3 \end{pmatrix}$ is the same as a single translation $\begin{pmatrix} 6 \\ -1 \end{pmatrix}$.

A to B to C: A rotation 90° clockwise about O followed by a reflection in the x-axis is the same as a single reflection in the line $y = -x$.

Worked example

target **C**

Triangle A is shown on the grid.

(a) Reflect triangle A in the y-axis. Label your new triangle B.
(1 mark)

(b) Triangle B is reflected in the line $y = 1$ to give triangle C. Describe fully the **single** transformation which takes triangle A onto triangle C.
(4 marks)

Rotation 180° about the point (0, 1)

To answer part **(b)** you need to draw
- the line $y = 1$
- the new triangle, C.

For a rotation of 180° you don't need to give a direction.

Check it!
You can ask for tracing paper in the exam.
Check a reflection by folding the tracing paper along the symmetry line. ✓

Describe fully...
A translation: vector of translation. ☑

A reflection: equation of mirror line. ☑

A rotation: angle of turn, direction of turn and centre of rotation. ☑

An enlargement: scale factor and centre of enlargement. ☑

Now try this

target **C**

Triangle A is shown on the grid.

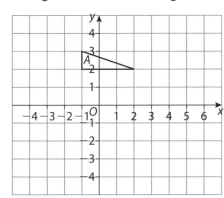

(a) Translate triangle A by the vector $\begin{pmatrix} 3 \\ -3 \end{pmatrix}$. Label the image B.
(2 marks)

(b) Rotate triangle B 90° clockwise about (0, 1). Label the image C.
(2 marks)

(c) Describe fully the **single** transformation that takes triangle C onto triangle A.
(3 marks)

Check a rotation by putting your pencil on the centre of rotation and turning the tracing paper. ✓

Similar shapes 2

The relationship between similar shapes is defined by a SCALE FACTOR.
A and B are similar shapes. B is an enlargement of A with scale factor k.

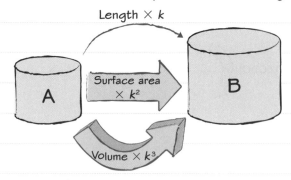

When a shape is enlarged by a linear scale factor k

- Enlarged surface area
  $= k^2 \times$ original surface area
- Enlarged volume
  $= k^3 \times$ original volume
- Enlarged mass $= k^3 \times$ original mass

Worked example

These two glass prisms are similar in shape.

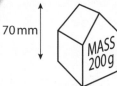 

70 mm

MASS 200 g

MASS 600 g

The 200 g prism is 70 mm high.
Work out the height of the 600 g prism.

(3 marks)

$200 \times k^3 = 600$

$k^3 = \dfrac{600}{200} = 3$

$k = \sqrt[3]{3} = 1.4422...$

Enlarged height $= 70 \times 1.4422...$
$= 101$ mm (3 s.f.)

EXAM ALERT!

Always use k, k^2 and k^3 to write the relationship.

Enlarged mass $= k^3 \times$ original mass

Solve the equation to find the value of k. Use the button on your calculator.

Multiply the original height by k to find the height of the enlarged shape.

Students have struggled with exam questions similar to this – **be prepared!**

Comparing volumes

You can use k^3 to compare volume, mass or capacity.

$k = \dfrac{32}{16} = 2$

Volume of large bottle
$= 1.2 \times k^3$
$= 1.2 \times 8$
$= 9.6$ litres

1.2 litres

←16 cm→   ←32 cm→

Now try this

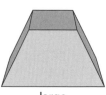

Here are three mathematically similar containers.
The table shows some information about these containers.

	Height (cm)	Area of top of container (cm²)	Volume (cm³)
small	12	85	Y
medium	36	X	7830
large	W	2125	Z

small
medium
large

Work out the missing values, W, X, Y and Z.
(6 marks)

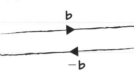

Vectors

A vector has a MAGNITUDE (or size) and a DIRECTION.

This vector can be written as **a**, $\overrightarrow{AB}$ or $\binom{2}{5}$.

You can multiply a vector by a number. The new vector has a different length but the same direction.

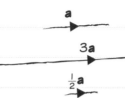

If **b** is a vector then −**b** is a vector with the same length but opposite direction.

Worked example

In the diagram, OADB and ACED are parallelograms. M is the midpoint of CE and D is the midpoint of BE.

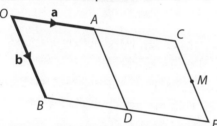

Express in terms of **a** and **b** the following vectors. Give your answers in simplest form.

(a) $\overrightarrow{OD}$ *(1 mark)*

$\overrightarrow{OD} = \overrightarrow{OA} + \overrightarrow{AD}$

$\quad = a + b$

(b) $\overrightarrow{BC}$ *(1 mark)*

$\overrightarrow{BC} = \overrightarrow{BD} + \overrightarrow{DE} + \overrightarrow{EC}$

$\quad = a + a - b = 2a - b$

(c) $\overrightarrow{OM}$ *(1 mark)*

$\overrightarrow{OM} = \overrightarrow{OA} + \overrightarrow{AC} + \frac{1}{2}\overrightarrow{CE}$

$\quad = a + a + \frac{1}{2}b = 2a + \frac{1}{2}b$

Adding vectors

You can add vectors using the TRIANGLE LAW. You trace a path along the added vectors to find the new vector.

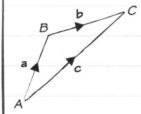

$a + b = c$

c is the resultant vector of **a** and **b**.

$A \rightarrow B \rightarrow C$ is the same as $A \rightarrow C$

For each vector, trace a path along the shape from the start point to the end point. If you go in the **opposite** direction to the vector then you need to **subtract**.

M is the midpoint of CE. This means that $\overrightarrow{CM} = \frac{1}{2}\overrightarrow{CE} = \frac{1}{2}b$.

Now try this

In triangle OAB,

M is the midpoint of OA, AN $= \frac{3}{4}$ AB and P and Q are points of trisection of OB.

$\overrightarrow{OM} = $ **a** and $\overrightarrow{OP} = $ **b**

Work out expressions for these vectors.

Give your answers in terms of **a** and **b**, in their simplest form.

(a) $\overrightarrow{OA}$ *(1 mark)* **(b)** $\overrightarrow{OB}$ *(1 mark)*

(c) $\overrightarrow{PM}$ *(1 mark)* **(d)** $\overrightarrow{AB}$ *(1 mark)*

(e) $\overrightarrow{ON}$ *(2 marks)* **(f)** $\overrightarrow{QN}$ *(2 marks)*

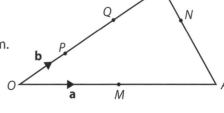

A* A B C D

Solving vector problems

Parallel vectors

If one vector can be written as a MULTIPLE of the other then the vectors are PARALLEL.

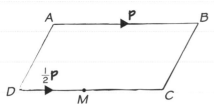

In this parallelogram M is the midpoint of DC. AB is parallel to DM so $\overrightarrow{DM} = \frac{1}{2}\overrightarrow{AB}$.

Remember that AB means the line segment AB (or the length of the line segment AB). $\overrightarrow{AB}$ means the vector which takes you from A to B.

Collinear points

If three points lie on the SAME STRAIGHT LINE then they are collinear. Here are three points:

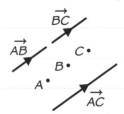

If any TWO of the vectors $\overrightarrow{AB}$, $\overrightarrow{BC}$ or $\overrightarrow{AC}$ are parallel, then the three points must be collinear.

Worked example

target A*

In the diagram, OPQ is a triangle. Point R lies on the line PQ such that PR : RQ = 1 : 2. Point S lies on the line through OR such that OR : OS = 1 : 3.

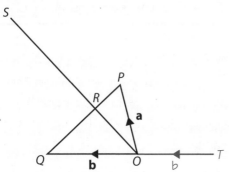

(a) Show that $\overrightarrow{OS} = 2\mathbf{a} + \mathbf{b}$.   (3 marks)

$\overrightarrow{PQ} = -\mathbf{a} + \mathbf{b}$, so $\overrightarrow{PR} = \frac{1}{3}(-\mathbf{a} + \mathbf{b})$

$\overrightarrow{OR} = \mathbf{a} + \frac{1}{3}(-\mathbf{a} + \mathbf{b})$

$\quad = \frac{1}{3}(2\mathbf{a} + \mathbf{b})$

$\overrightarrow{OS} = 3\,\overrightarrow{OR}$

$\quad = 2\mathbf{a} + \mathbf{b}$

(b) Point T is added to the diagram such that $\overrightarrow{TO} = \mathbf{b}$. Prove that points T, P and S lie on the same straight line. (3 marks)

$\overrightarrow{TP} = \mathbf{a} + \mathbf{b}$

$\overrightarrow{TS} = \mathbf{b} + 2\mathbf{a} + \mathbf{b}$

$\quad = 2\mathbf{a} + 2\mathbf{b}$

$\quad = 2(\mathbf{a} + \mathbf{b})$

So $\overrightarrow{TS}$ and $\overrightarrow{TP}$ are parallel, and T, S and P are collinear.

You might be given information about the lengths of lines as ratios.

PR : RQ = 1 : 2. There are 2 + 1 = 3 parts in this ratio. This means that R is $\frac{1}{3}$ of the way along PQ so $\overrightarrow{PR} = \frac{1}{3}\overrightarrow{PQ}$.

OR : OS = 1 : 3. This means that $\overrightarrow{OS} = 3\,\overrightarrow{OR}$.

If the question tells you another point has been added to a diagram you should always draw it on your exam paper. To show that T, P and S lie on the same straight line you need to show that **two** of the vectors $\overrightarrow{TP}$, $\overrightarrow{TS}$ and $\overrightarrow{PS}$ are parallel.

Now try this

target A*

In triangle OAB, M is the midpoint of OA.

OH = HJ = JK = KB

S and T are points of trisection of AB.

$\overrightarrow{OM} = \mathbf{a}$ and $\overrightarrow{OH} = \mathbf{b}$

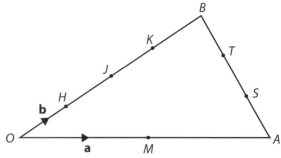

(a) Prove that HA, JS and KT are all parallel.
(6 marks)

(b) State the ratio KT : JS : HA   (1 mark)

Problem-solving practice 1

About half of the questions on your exam will need problem-solving skills.

These skills are sometimes called AO2 and AO3.

Practise using the questions on the next two pages.

For these questions you might need to:

- choose which mathematical technique or skill to use
- apply a technique in a new context
- plan your strategy to solve a longer problem
- show your working clearly and give reasons for your answers.

---

1 *

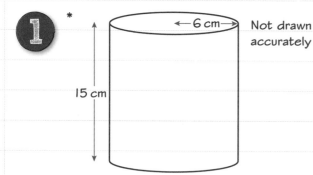
←—6 cm—→ Not drawn accurately

15 cm

Jenny fills some empty flowerpots completely with compost.

Each flowerpot is in the shape of a cylinder of height 15 cm and radius 6 cm.

Jenny has a 15 litre bag of compost.

She fills up each flowerpot completely.

How many flowerpots can she fill completely?

You must show your working.   *(4 marks)*

Circles and cylinders p.85

You need to remember that 1 litre = 1000 cm^3.

You are trying to work out how many flowerpots Jenny can fill **completely** so you'll need to round your final answer **down**.

TOP TIP

If a question has a * next to it, then there are marks available for QUALITY OF WRITTEN COMMUNICATION. This means you must show all your working and write your answer clearly with the correct units.

---

2 A ladder is 6 m long.

The ladder is placed on horizontal ground, resting against a vertical wall.

The instructions for using the ladder say that the bottom of the ladder must not be closer than 1.5 m to the bottom of the wall.

How far up the wall can the ladder reach if the instructions are followed?

    (3 marks)

Pythagoras' theorem p.78

You should definitely draw a sketch to show the information in the question.

TOP TIP

Be careful when you are working out the length of a **short** side using Pythagoras' theorem.

Remember: short^2 + short^2 = long^2

short^2 = long^2 − short^2

Problem-solving practice 2

3

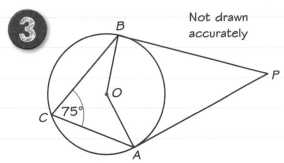

Not drawn accurately

In the diagram A, B and C are points on the circumference of a circle with centre O.

PA and PB are tangents to the circle.

Angle ACB = 75°.

Prove that angle APB is 30°.

(4 marks)

Circle theorems p.74
Circle facts p.73

You have to write down a reason for **each step** of your working. These are some of the reasons you could use to answer this question:

- Angle at the centre of a circle is twice the angle at the circumference.
- Angle between a tangent and a radius is 90°.
- Angles in a quadrilateral add up to 360°.

TOP TIP

When you are learning circle theorems, draw a sketch to explain each one. This will help you to spot which theorem to use in an exam question.

4

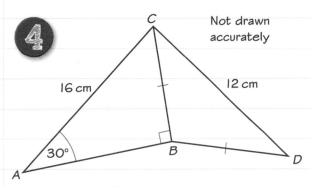

Not drawn accurately

AC = 16 cm. CD = 12 cm.

BC = BD.

Angle ABC = 90°. Angle CAB = 30°.

Work out the area of triangle BCD.

(6 marks)

The cosine rule p.83
Triangles and segments p.87

You need more information about triangle BCD before you can calculate its area.

Triangle ABC is right angled so you can use $S^O_H C^A_H T^O_A$ to work out the length of BC. Then use the cosine rule to work out the size of angle BCD. Finally, you can use the formula $A = \frac{1}{2}ab \sin C$ to work out the area of triangle BCD.

TOP TIP

Write down any values from your calculator to at least 4 decimal places before rounding your final answer to the required degree of accuracy.

---

5 The graph of the curve $y = a + b^x$ passes through points (1, 9) and (2, 13), where $a > 0$ and $b > 0$.

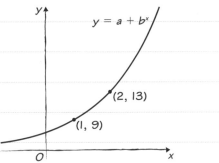

Work out the values of a and b. **(6 marks)**

Graphs of curves p.65
Quadratic equations p.49

Substitute the values given into the equation of the graph:

$9 = a + b^1$ (1)

$13 = a + b^2$ (2)

You can use (1) to write $a = 9 - b$ and substitute this into (2). You need to solve a quadratic equation to find the value of b.

TOP TIP

Every point on a graph satisfies the equation of the graph.

Answers

UNIT 1

1. Calculator skills 1

1 529 boys   **2** 76.8%

2. Percentage change 1

1 £14 310   **2** 34.6%

3. Reverse percentages and compound interest

1 £45   **2** 202 000 (3 s.f.)

4. Ratio

1 112 g   **2** 18.75%

5. Standard form

1 **(a)** 2.8×10^{-4} **2** **(a)** 2.71×10^{6}
   (b) 391 000   **(b)** 5.53×10^{11}

6. Upper and lower bounds

1 81.4 mph (1 d.p.)
2 6.6 cm (unrounded answer = 6.5996…)

7. The Data Handling Cycle

Select, at random, 20 girls and 20 boys from your year group, and ask them to keep a record, over the next week, of how many times they log on to a social networking site.

Calculate an average (mean and/or median) for each group.

Write a conclusion making a statement about 'average usage' for both girls and boys and say whether the hypothesis is likely to be true or false.

8. Collecting data

(a) Secondary, Discrete   **(b)** Secondary, Continuous
(c) Primary, Discrete

9. Surveys

(a)

	Tally	Frequency
Buy lunch		
Bring from home		

(b) How much do you spend on lunch each day?
£0–£0.99 ☐   £1–£1.99 ☐   £2–£2.99 ☐   £3–£3.99 ☐
£4 or more ☐

10. Two-way tables

	Boys	Girls	Total
Left-handed	3	6	9
Right-handed	11	12	23
Total	14	18	32

11. Sampling

Age group	Junior	18–39	40–59	Senior
Number of members	120	180	270	150
Number in sample	20	30	45	25

12. Mean, median and mode

19

13. Frequency table averages

42.25 years

14. Interquartile range

(a)

```
0 | 8
1 | 0 2 2 4 4 5 6 6 6 8 9 9
2 | 0 0 1 2 3 3 6 8
3 | 1 2 4 6
4 | 0 5
```

Key 2|1 represents 21 lessons

(b) Median = 20   IQR = 28 − 15 = 13

15. Scatter graphs

(a), (b)

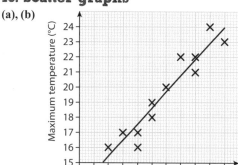

(c) Positive
(d) The greater the number of hours of sunshine, the greater the maximum temperature. There is a strong, positive correlation.
or,
As the number of hours of sunshine increased, so did the maximum temperature. There is a strong, positive correlation.

16. Frequency polygons

(a)

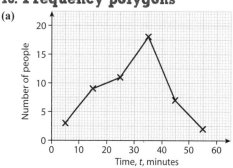

(b) $30 < t \leqslant 40$

17. Histograms

(a)

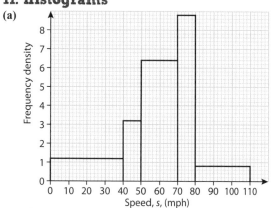

(b) $\dfrac{(\frac{128}{4} + 88 + 24)}{320} \times 100 = \dfrac{144}{320} \times 100 = 45\%$

18. Probability 1
(a) 0.6   **(b)** 0.9   **(c)** $0.6 \times 0.6 = 0.36$

19. Probability 2
(a) $420 \times 0.15 + 390 \times 0.2 = 63 + 78 = 141$

(b) $\frac{141}{810} = 0.174$

20. Tree diagrams
$\frac{70}{132} = \frac{35}{66}$

21. Cumulative frequency
(a) 29 minutes   **(b)** $38 - 22 = 16$ minutes

(c) $500 - 430 = 70, \frac{70}{500} \times 100 = 14\%$

22. Box plots
(a)

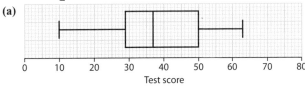

(b) 30 boys   **(c)** approximately 15 boys

23. Comparing data
(a)

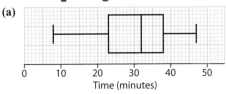

(b) 1st comment: On average, the boys completed the challenge in a faster time than the girls (median: boys 29 mins, girls 32 mins).

2nd comment: The girls' times were more closely grouped than the boys' times. (IQR: girls 15 mins, boys 24 mins or range: girls 39 mins, boys 44 mins).

24–25. Problem-solving practice 1–2
1 £223.56 for Business First, £208 for Leisure One … so 'No'.

2 0.33 seconds (to 2 d.p.)

3 312 students

4

School	Number of students	Number in sample
Bounds green	258	17
Chestnuts	394	26
Mulberry	679	45
Campsbourne	182	12

5 3 possible successful outcomes $= \frac{3}{30} = \frac{1}{10}$

26. Factors and primes
(a) $2 \times 2 \times 3 \times 3 \times 5$ **(b)** 90   **(c)** 1260

27. Fractions and decimals
1 $\frac{9}{20}$, 0.6, 65%   **2** $\frac{37}{40}$ **3** $\frac{49}{20}$ or $2\frac{9}{20}$

28. Decimals and estimation
1 (a) 3.813   **(b)** 155   **(c)** 10

2 4000

29. Recurring decimals
1 $\frac{7}{12}$ because 12 has a prime factor of 3.

2 0.4545… (minimum of 4 d.p.) or $0.\dot{4}\dot{5}$

30. Percentage change 2
Cruks (£171.50 as against £175.50 at Spivs)

31. Ratio problems
$288 : 192 = 3 : 2$

32. Indices 1
1 (a) 80   **(b)** 4 and -4

2 (a) 12   **(b)** 1   **(c)** x^{12} **(d)** y^{15}

33. Indices 2
1 (a) 7^{h-k} **(b)** 7^{2h} **(c)** 7^{h+2k}

2 $\frac{1}{27}$

3 $32 \div \frac{1}{8} = 256$

34. Surds
1 (a) $11\sqrt{2}$ **(b)** $5\sqrt{7}$

2 32

35. Algebraic expressions
1 h^{12}

2 (a) $16a^{20}b^4$ **(b)** $15x^7y^9$ **(c)** $3d^6g^5$

3 $5p^3$

36. Expanding brackets
1 $5a - 30$

2 (a) $2b - 11$ **(b)** $24y^4 + 32y$
   (c) $2x^3 + 3x^2 - 3x$ **(d)** $m^2 + 6m - 27$

3 (a) $6g^2 - 23ge + 20e^2$ **(b)** $16x^2 + 56x + 49$

37. Factorising
1 (a) $2(2a - 3)$ **(b)** $y(y + 5)$

2 (a) $3g(4 + g)$ **(b)** $(p - 14)(p - 1)$
   (c) $2x(3x - 4y)$

3 (a) $4ma(1 - 6m)$ **(b)** $(p + 8)(p - 8)$

4 (a) $(3y - 5)(y + 4)$ **(b)** $(x + 9y)(x - 9y)$
   (c) $2(5g + e)(5g - e)$

38. Algebraic fractions
1 (a) $\frac{2a + 9}{6a}$ **(b)** $\frac{1}{y - 3}$

2 (a) $\frac{x(x - 2)}{x + 5}$ **(b)** $\frac{m - 6}{3m^2}$

39. Straight-line graphs
(a) y

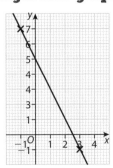

(b) Gradient $= -2$, y-intercept $= 5$

40. Gradients and midpoints
$y = 5x - 20$

41. Real-life graphs
(a) 104 km   **(b)** 45 minutes   **(c)** 112 km/h

42. Formulae

(a) $C = 10d + 30$

(b)

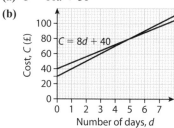

(c) 5 days

43. Linear equations 1

1 **(a)** $w = 7$ **(b)** $x = -1$

2 **(c)** $y = -\frac{7}{4}$ **(d)** $m = 4$

44. Linear equations 2

1 **(a)** $w = -5$ **(b)** $x = 3$

2 **(a)** $y = 6$ **(b)** $m = 37$

45. Number machines

-3.5

46. Inequalities

1 $-2, -1, 0, 1, 2$

2 $-4, -3, -2, -1, 0$

3 $n > -2$

4 $n \geqslant 22$

47. Inequalities on graphs

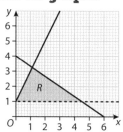

48. Simultaneous equations 1

1 $x = 5, y = 1.5$ **2** $x = 4, y = -1$

3 pen = 12p, ruler = 19p

49. Quadratic equations

1 $m = 2, m = 6$ **2** $w = -4, w = 9$

3 $y = \frac{3}{5}, y = -8$ **4** $x = \frac{5}{7}, x = -1$

50. Quadratics and fractions

1 $x = \frac{3}{4}, x = -3$

2 $x = -\frac{15}{2}, x = -4$

51. Simultaneous equations 2

1 $x = 5, y = 1$ and $x = -3, y = -3$

2 $x = -\frac{2}{5}, y = -\frac{9}{5}$ and $x = 3, y = 5$

52. Rearranging formulae

1 $t = \dfrac{4p + 1}{3}$ **2** $w = \dfrac{m^2 - 7}{5}$

3 $y = \dfrac{5 - 2x}{x + 9}$

53. Sequences 1

(a) $3n - 1$ **(b)** 86 **(c)** 66

(d) No, 52nd term = 155, 53rd term = 158

54. Sequences 2

1 **(a)** 31, 35, 39 **(b)** 19th term

2 $2(b + 9) = 14 \rightarrow b = -2$ $2(a + 9) = -2 \rightarrow a = -10$

55. Algebraic proof

1 *Example answer:* $a = 3, b = 2, (2a + b)^2 = 8^2 = 64$ which is even.

2 $(n + 4)^2 - n^2 = n^2 + 8n + 16 - n^2$
$$= 8n + 16$$
$$= 8(n + 2)$$

The mean of $(n + 4)$ and $n = \dfrac{n + 4 + n}{2} = n + 2$

Therefore, the difference between the squares, $8(n + 2)$, is eight times the mean, $(n + 2)$

56. Identities

$h = 15, k = -5$ and $h = -15, k = 5$

57. Completing the square

1 $x = -5 \pm \sqrt{13}$

2 $y = 3 \pm 2\sqrt{6}$

58–59. Problem-solving practice 1–2

1 40 at £4.50 + 35 at £3.00 = £285.00

2 Amir is 7 years old.

3 $k = 11$

4 $C(0, 3), D(-6, 0)$

5 $a = 9$

6 $(n + 1)^2 - n^2 = (n^2 + 2n + 1) - n^2$
$$= 2n + 1$$
$$= n + (n + 1)$$

60. Proportion

1 £76.90 **2** 16 days

61. Trial and improvement

1 $x = 2.8$ **2** $x = 3.4$

62. The quadratic formula

1 1.39 and -2.89 **2** 2.19 and -1.52

63. Quadratic graphs

(a)

x	-2	-1	0	1	2	3	4
y	4	-1	-4	-5	-4	-1	4

(b), (c)

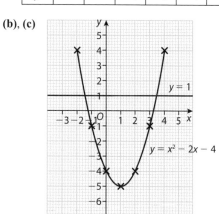

(d) From $(-1.5, 1)$ to $(-1.4, 1)$ **and** from $(3.4, 1)$ to $(3.5, 1)$

64. Using quadratic graphs

(a)

x	-5	-4	-3	-2	-1	0	1	2
y	7	1	-3	-5	-5	-3	1	7

(b)

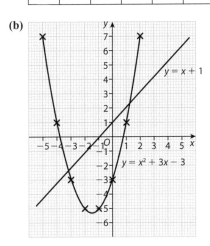

(c) -3.8 and 0.8

(d) $y = x + 1$, $x = 1.2$ and -3.2

65. Graphs of curves

(a)

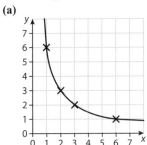

(b) $(\sqrt{6}, \sqrt{6})$

66. 3-D coordinates

(a) $(4, 0, 0)$ **(b)** $(0, 0, 5)$ **(c)** $(4, 9, 5)$
(d) $(4, 4.5, 5)$ **(e)** $(2, 4.5, 2.5)$

67. Proportionality formulae

(a) $P = \dfrac{7200}{Q}$

(b)

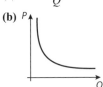

(c) $Q = 60$

68. Transformations 1

(a), (b)

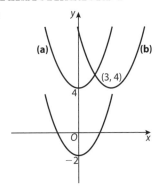

(c) 1st translation $y = x^2 - 2 \rightarrow y = x^2 + 4$

2nd translation
$y = x^2 + 4 \rightarrow y = (x - 3)^2 + 4 \rightarrow y = x^2 - 6x + 13$

69. Transformations 2

(a) $y = 2f(x)$

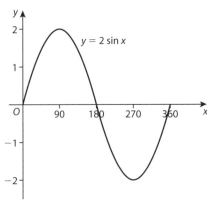

(b) $y = f(2x)$

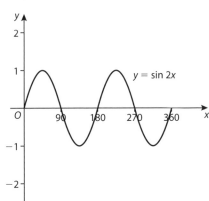

(c) $y = f(x - 90°)$

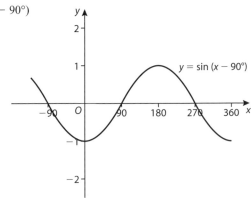

70. Angle properties

$x = 59°$

71. Solving angle problems

angle $DCB = \dfrac{180 - 2x}{2} = 90 - x$ (base angles isos $\triangle BDC$)

angle $ABC = 90 - x$ (base angles isos $\triangle ABC$)

angle $BAC = 180 - (90 - x) - (90 - x)$ (angle sum of $\triangle ABC$)
$\qquad\qquad = 180 - 90 + x - 90 + x$
$\qquad\qquad = 2x$

72. Angles in polygons

$x = 144°$

73. Circle facts

$QO = 7\,\text{cm}$ and $TQ = 16\,\text{cm}$

$TO^2 = 16^2 + 7^2 \rightarrow TO = 17.46...$ (Pythagoras' theorem)

$TW = TO - WO = 17.46... - 7 = 10.46...\,\text{cm}$ or $10.5\,\text{cm}$

74. Circle theorems

angle BAC = angle CBQ = x (alternate segment theorem)
angle BAD = $180° - 116° = 64°$ (opposite angles of cyclic quad.
$x + x + 8 = 64$ add up to $180°$)
        $x = 28°$

75. Perimeter and area

(a) $172 \, cm^2$ **(b)** $51 \, cm$

76. Similar shapes 1

(a) $130°$ **(b)** $31.5 \, cm$

77. Congruent triangle proof

In triangles BHC and CKB,
BC is common
angle HBC = angle KCB (base angles isosceles triangle ABC)
$BH = AB - AH$
$CK = AC - AK$
but $AB = AC$ and $AH = AK$ (given)
so $BH = CK$

So, triangles BHC and CKB are congruent   (SAS)

78. Pythagoras' theorem

(a) $y = 3.5 \, cm$ **(b)** $z = 9.1 \, cm$

79. Pythagoras in 3-D

(a) $TW = 18.4 \, cm$ (1 d.p.)
(b) $WX > YX$ since WX is the hypotenuse of triangle WYX.
    In $\triangle TWX$, T rises 7 cm in a horizontal distance of WX.
    In $\triangle TYX$, T rises 7 cm in a horizontal distance of YX.
    Since $WX > YX$, T rises 7 cm in a shorter distance (YX) in
    triangle TYX so angle TYX is larger than angle TWX.

80. Trigonometry 1

(a) $44.8°$ **(b)** $56.2°$ **(c)** $48.6°$

81. Trigonometry 2

(a) $4.4 \, cm$ **(b)** $5.4 \, cm$ **(c)** $15.7 \, cm$

82. The sine rule

(a) $45.2°$ (3 s.f.)   **(b)** $15.2 \, cm$ (3 s.f.)

83. The cosine rule

Angle $PQR = 39.8°$ (3 s.f.)

84. Trigonometry in 3-D

Angle $MJL = 25.0°$ (3 s.f.)

85. Circles and cylinders

$119.27 \, cm^2$ or $119.3 \, cm^2$ or $119 \, cm^2$ or $120 \, cm^2$

86. Sectors of circles

Angle $AOB = 129°$

87. Triangles and segments

Area sector $ABC = \dfrac{135°}{360°} \times \pi \times 4^2 = \dfrac{3}{8} \times \pi \times 16 = 6\pi \, cm^2$

Area $\triangle AOC = \dfrac{1}{2} \times 4 \times 4 \times \sin 135° = \dfrac{8}{\sqrt{2}} = \dfrac{8\sqrt{2}}{2} = 4\sqrt{2} \, cm^2$

Area minor segment ABC = Area sector − area $\triangle = 6\pi - 4\sqrt{2} \, cm^2$

88. Prisms

Area of cross-section = $\left(15 \times 3 + \dfrac{1}{2} \times 10 \times 4\right) cm^2 = 65 \, cm^2$
Volume of shape = $65 \times 8 = 520 \, cm^3$

89. Volumes of 3-D shapes

1 $r = 4 \, cm$
2 Volume box = $\pi \times (3.5)^2 \times 7$
    Volume ball = $\dfrac{4}{3} \times \pi \times (3.5)^3$
    Fraction = $\dfrac{4 \times \pi \times (3.5)^3}{3 \times \pi \times (3.5)^2 \times 7} = \dfrac{2}{3}$

90. Surface area

Slant height = $\sqrt{15^2 + 8^2} = 17$
Area cross-section = $\dfrac{1}{2} \times 16 \times 15 = 120 \, cm^2$
Area base = $\dfrac{1}{2} \times \pi \times 8^2 = 100.5309\ldots \, cm^2$
Area curved surface = $\dfrac{1}{2} \times \pi \times 8 \times 17 = 213.6283\ldots \, cm^2$
Total surface area = $120 + 100.5309\ldots + 213.6283\ldots$
                 = $374.159\ldots = 434 \, cm^2$ (3 s.f.)

91. Plans and elevations

(a)   **(b)**   **(c)**

(d) $30 \, cm^2$

92. Bearings

(a) $275 \, km - 290 \, km$ **(b)** $223° - 227°$
(c)
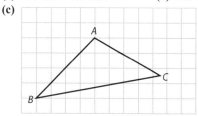

93. Scale drawings and maps

(a) $9 \, km$ **(b)** $16 \, cm$

94. Constructions

(a), (b)
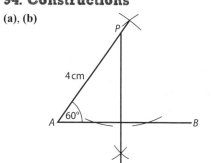

95. Loci

(a) (b) (i) (ii)
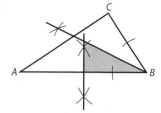

96. Speed

1 4 hours 18 minutes   **2** 264 miles

97. Density

1 624 g

2 10.49 g/cm³

98. Converting units

1 81.25 mph

2 12 gal = 54 litres, cost = £71.28

3 111.6 km/h

99. Translations, reflections and rotations

(a) Rotation, 90° clockwise, about (2, 2)

(b) $\begin{pmatrix} -5 \\ -2 \end{pmatrix}$

100. Enlargements

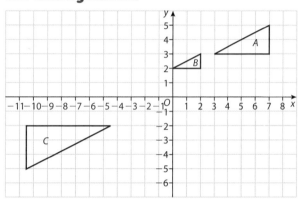

101. Combining transformations

(a), (b)

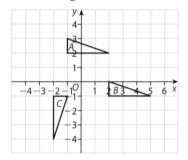

(c) Rotation, 90° anti-clockwise about (−3, 1).

102. Similar shapes 2

$W = 60$ cm $X = 765$ cm² $Y = 290$ cm³ $Z = 36\,250$ cm³

103. Vectors

(a) $2a$

(b) $3b$

(c) $-b + a$

(d) $-2a + 3b$

(e) $\frac{1}{2}a + \frac{9}{4}b$

(f) $\frac{1}{2}a + \frac{1}{4}b$

104. Solving vector problems

(a) $AB = -2a + 4b$

$\overrightarrow{HA} = -b + 2a$ or $2a - b$

$\overrightarrow{JS} = 2b - \frac{2}{3}(-2a + 4b)$

$\qquad = 2b + \frac{4a}{3} - \frac{8b}{3} = \frac{4a}{3} - \frac{2b}{3} = \frac{2}{3}(2a - b)$

$\overrightarrow{KT} = b - \frac{1}{3}(-2a + 4b)$

$\qquad = b + \frac{2a}{3} - \frac{4b}{3} = \frac{2a}{3} - \frac{1b}{3} = \frac{1}{3}(2a - b)$

HA, JS and KT are all multiples of $(2a - b)$, so are parallel.

(b) $KT : JS : HA = 1 : 2 : 3$

105–106. Problem-solving practice 1–2

1 8 flowerpots

2 5.81 m (3 s.f.)

3 Angle $AOB = 150°$ (Angle at the centre of a circle is twice the angle at the circumference)

Angle $OAP =$ Angle $OBP = 90°$ (Angle between a tangent and a radius is 90°)

Angle $AOB +$ Angle $OAP +$ Angle $OBP +$ Angle $APB = 360°$ (Angles in a quadrilateral add up to 360°)

So 150° + 90° + 90° + Angle $APB = 360°$

So Angle $APB = 30°$

4 31.7 cm² (3 s.f.)

5 $a = \dfrac{17 - \sqrt{17}}{2} = 6.44$ (3 s.f.), $b = \dfrac{1 + \sqrt{17}}{2} = 2.56$ (3 s.f.)

Published by Pearson Education Limited, Edinburgh Gate, Harlow, Essex, CM20 2JE.

www.pearsonschoolsandfecolleges.co.uk

Text and original illustrations © Harry Smith and Pearson Education Limited 2013
Edited and produced by Wearset, Boldon, Tyne and Wear
Typeset and illustrated by Tech-Set Ltd, Gateshead
Cover illustration by Miriam Sturdee

The right of Harry Smith to be identified as author of this work has been asserted by him in accordance with the Copyright, Designs and Patents Act 1988.

First published 2013

16 15 14 13 12
10 9 8 7 6 5 4 3 2

British Library Cataloguing in Publication Data
A catalogue record for this book is available from the British Library

ISBN 978 1 447 94136 1

Printed in Slovakia by Neografia

Every effort has been made to contact copyright holders of material reproduced in this book. Any omissions will be rectified in subsequent printings if notice is given to the publishers.

In the writing of this book, no AQA examiners authored sections relevant to examination papers for which they have responsibility.

There are no questions printed on this page.

There are no questions printed on this page.

There are no questions printed on this page.

There are no questions printed on this page.

Revision is more than just this Guide!

You'll need plenty of practice on each topic you revise

1-to-1 page match with this Revision Guide.

Guided questions help build your confidence.

Exam-style questions on this topic.

Grades, marks and hints get you well prepared for this topic in your exam.

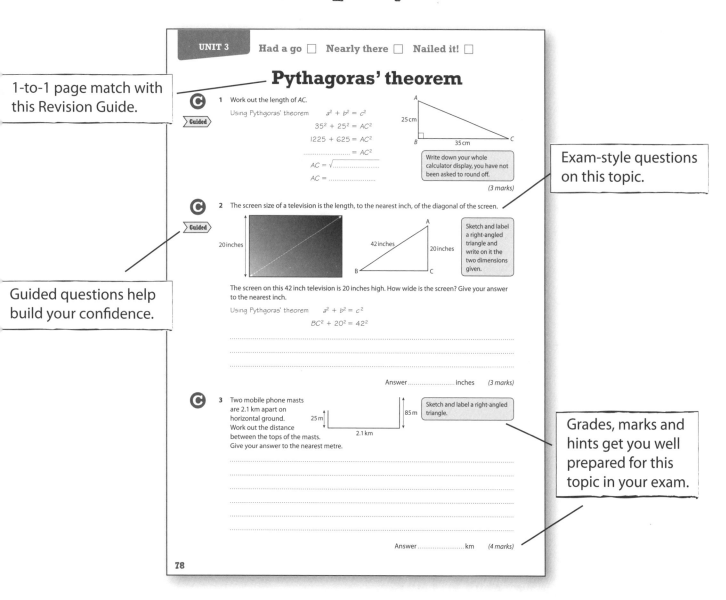

Check out the matching Workbook!

Revision is more than just this Guide!